The Complexity of Robot Motion Planning

ACM Doctoral Dissertation Awards

1982

Area-Efficient VLSI Computation
by Charles Eric Leiserson

1983

Generating Language-Based Environments
by Thomas W. Reps

1984

Reduced Instruction Set Computer Architectures for VLSI
by Manolis G.H. Katevenis

1985

Bulldog: A Compiler for VLIW Architectures
by John R. Ellis

1986

Computational Limitations of Small Depth Circuits
by Johann Håstad

Full Abstraction and Semantic Equivalence
by Ketan Mulmuley

1987

The Complexity of Robot Motion Planning
by John F. Canny

The Complexity of Robot Motion Planning

John Canny

The MIT Press
Cambridge, Massachusetts
London, England

Printed and bound in the United States of America.

Library of Congress Cataloging-in-Publication Data

Canny, John.

The complexity of robot motion planning / John Canny.

p. cm. -- (ACM doctoral dissertation awards ; 1987)

Bibliography: p.

Includes index

ISBN 0-262-03136-1

1. Robots--Motion. I. Title. II. Series: ACM doctoral dissertation award: 1987.

T J211.4.C36 1988

629.8'92--dc 19

In memory of Pat Canny

Contents

List of Figures

Series Foreword

The ACM Doctoral Dissertation Award is presented each year by the Association for Computing Machinery (ACM). This award recognizes the best English-language dissertation written in a computer-related field during the academic year. The winning author receives a cash award of $1,000 from ACM. In addition, the winning dissertation is published by the MIT Press, with the author receiving appropriate royalties.

This was the sixth competition to be held. Schools were asked to nominate their best dissertations among those accepted between July 1, 1986, and June 30, 1987. By a screening process in which each of the forty-one nominated dissertations was reviewed by several external referees, five dissertations were selected for final review. A committee with representatives from both industry and academia then read the five final selections and chose the winner. The members of the committee were Doug Degroot, David Johnson, Fred Maryanski, Jack Minker, Steve Muchnik, and Larry Snyder. The dissertation they chose was "The Complexity of Robot Motion Planning," by John F. Canny. Dr. Canny's thesis work was supervised by Professor Tomás Lozano-Pérez of the Massachusetts Institute of Technology.

Canny's thesis is a ground-breaking study of the algorithmic aspects of robotics, with material enough for two award winners. In particular he is concerned with the problem of planning 3-dimensional movement in the presence of obstacles and other constraints. He resolves long-open problems concerning the complexity of motion planning problems, and for the central problem of finding a collision free path for a jointed robot in the presence of obstacles, he obtains exponential speedups over existing algorithms by applying high-powered new mathematical techniques. Moreover, he has paid much attention to his representations and algorithms, so that both the high level mathematics and the low level implementation details are presented with unusual precision and clarity. The work presents a truly outstanding example of doctoral research in computer science.

David S. Johnson, Chairman
ACM Doctoral Dissertation Award Subcommittee

Preface

Robotics is a discipline that lies on the boundary between several fields. In this thesis we are interested in the computational complexity of planning collision-free motions for a robot in an environment filled with obstacles. The tools for analysis of algorithms and complexity come from computer science, but we also need ideas from differential topology and commutative algebra. The results we obtain depend crucially on the use of all of these ideas, and would be impossible without the same level of cross-fertilization.

The main problem we tackle is the "generalized movers' problem", which is the problem of moving a general robot, which could be an arm or a mobile robot, in a three dimensional environment filled with obstacles. The first general solution for this problem was given by Schwartz and Sharir [SS], who showed that it could be solved in time double-exponential in the number of degrees of freedom. Since then, there have been steady improvements in algorithms for special cases of the problem, and these often have exponents that are linear in the number of degrees of freedom. However, they involve enormous constants due to the cost of exact computation on algebraic numbers. The main contribution of this thesis is the roadmap algorithm, which solves the generalized movers' problem in single-exponential time, with exponent equal to the number of degrees of freedom. This algorithm equals or betters the asymptotic performance of most special purpose algorithms, and has dramatically lower constants.

The improvement derives from the following ideas: Firstly, we use a coarse partition or *stratification* of the configuration space of the robot. This stratification divides configuration space into a small number of simple geometric pieces. Secondly, we use *multivariate resultants* for solution of system of polynomials in many variables. The multivariate resultant allows elimination of several variables in single-exponential time. Finally, from bounds on the size of the multivariate resultant, we derive a very general gap theorem for systems of polynomials. The gap theorem gives lower bounds on the separation of distinct algebraic numbers defined by any system of polynomials of a certain size. The theorem makes it possible to use binary approximations to algebraic numbers with no loss of accuracy.

The roadmap algorithm can either be used directly on the set of collision-free configurations, or on a subset of maximal clearance configurations. This suggests the use of Voronoi diagrams, but the usual definition is very difficult to work with in a high-dimensional configuration space. Instead we introduce a new kind of Voronoi diagram which is based on a distance function derived directly from the boundary surfaces in configuration space. Its algebraic complexity is lower in many cases, and it is shown to be complete for motion planning.

We also give lower bounds for some extensions of the generalized movers' problem. While the movers' problem is solvable in polynomial time for any fixed number of degrees of freedom, when a shortest path is sought, or when the obstacles can move faster than the robot, the planning problem is provably hard even in three dimensions. In particular we show that the euclidean shortest path problem in three dimensions, a major open problem in computational geometry, is NP-hard. The "2-d asteroid avoidance problem" is also shown to be NP-hard. Finally, when compliant (sliding) motion of the robot is allowed, and uncertainty is taken into account, motion planning in three dimensions becomes hard for non-deterministic exponential time.

The thesis contains several chapters with fairly heavy mathematical content. For many of the geometric ideas, figures have been included. These should give the general reader a good flavor of what is going on. The dependencies between chapters have been kept to a minimum, and each can be read in isolation. The exception to this is chapter 1, which contains definitions and terminology used throughout the later chapters. Chapter 2 concerns computation of configuration space descriptions from physical descriptions of the robot and obstacles. It contains a short summary of the properties of quaternions, which are used to represent the configuration of a free polyhedron. Chapter 3 describes the multivariate resultant and the gap theorem. Chapter 4 is the heart of the thesis. Stratified sets are introduced, and the roadmap algorithm is described. Some familiarity with singularity theory is useful here. Chapter 5 describes the simplified Voronoi diagram, and uses simple homotopy arguments to show that the diagram is complete for motion planning. Chapter 6 is pure computational geometry. It gives lower bounds for problems of 3-d shortest path, 2-d asteroid avoidance, and 3-d complaint motion planning with uncertainty.

Acknowledgements

My sincere thanks to Prof. Tomás Lozano-Pérez, my thesis advisor at MIT for all his support, understanding and perspective throughout the work on this thesis. In addition to Tomás, I was lucky enough to be able to interact with some other truly exceptional individuals. John Reif taught me a geat deal about complexity theory in robotics and nature, and David Mumford gave me invaluable direction in the search for the right kind of mathematics to apply in this new field.

Thanks to Bruce Donald and Mike Erdmann for detailed comments on the thesis and for sharing the learning process. This version has greatly benefited from comments from Ken Goldberg, Barbara Moore and Doug Ierardi. Finally thanks to Stuart Russell, Robert Wilensky and the people in BAIR for providing, and keeping together, the text processing resources I needed.

The Complexity of Robot Motion Planning

Chapter 1

Introduction

The objective of robot motion planning is to automatically guide a robot around in an environment filled with obstacles, and to choose forces to be applied when assembling objects. Economic considerations often mandate the use of a cost function on the robot's motion, such as time duration or the amount of energy required. When planning a *trajectory* for the robot, which gives position as a function of time, planning must take into account physical limitations of the robot. For example, the motors that drive the robots's joints will have limited torque. The sampling rate of the digital controllers for the motors usually limits the maximum speed of a trajectory that the robot can track successfully. So there are many possible variations on the motion planning problem, where some or all of these extra constraints are taken into account. This thesis studies the complexity of several fundamental classes of problem, and introduces some novel geometric techniques for solving them.

The most difficult planning problems arise in assembly, where clearances between parts are typically smaller than the robot's positioning or sensing accuracy. The robot cannot consistently execute motions below this accuracy, and when it attempts to do so, the results of its actions are no longer deterministic. Such a motion can lead to any of several qualitatively different situations. There is no obvious alternative open to the planner except the enumeration of all the possibilities, leading to a combinatorial explosion with the number of plan steps.

The physics of contact, including the effects of friction which may also be nondeterministic, must be taken into account. To cope with this nondeterminism, assembly plans usually include sensing steps and allow branching. This allows the runtime system to distinguish and proceed from any of the several possible outcomes of some action. The sequence of execute/sense steps may be viewed as a "game with nature" [TMG], where the planner is trying to produce a strategy which will complete the assembly no matter what nature decides will be the outcomes of its actions.

The problems studied here include the most fundamental motion planning problem, the "generalized movers' problem". [1] We present an algorithm for this problem which runs in single exponential time. The only other general solutions to the problem require double exponential time. Compared to the many algorithms for special cases of the movers' problem, the new algorithm almost always has an equal or lower runtime exponent, with dramatically lower constants. The algorithm is called the "roadmap algorithm" since it generates a one-dimensional subset or "roadmap" of the possible motions of a robot. The exponent of the algorithm is in fact equal to the number of degrees of freedom. So while it is exponential in the worst case, most robots have a small, fixed number of degrees of freedom, and then the algorithm runs in low polynomial time.

The techniques used in the algorithm are straightforward and widely applicable. Hopefully they will help to open up the area of non-linear computational geometry, especially in higher dimensions. Work in this area so far has lead to some very intricate and complicated algorithms, with frustratingly high algebraic constants. By contrast, the roadmap has a one-line recursive definition, given in chapter 4. The class of geometric problems that can be tackled with these methods includes non-linear optimization problems, geometric theorem proving, kinematic analysis of robots, and static and dynamic geometric modeling.

The *generalized movers' problem* is the problem of planning an obstacle-avoiding *path* for a robot with any number of degrees of freedom. The generalized movers' problem is really a broad class of problems, since it includes planning problems for robots with arbitrary kinematics and number of degrees of freedom, and even includes coordinating the motion of several

[1] The nomenclature is due to Reif [Rei]

robots. With only two exceptions, all the papers on the movers' problem have considered very restricted classes of robots. There has been a considerable number of papers analysing these special cases, and survey articles are given in [Tou] and [SY]. A collection of algorithmic papers is given in the book by Hopcroft, Schwartz and Sharir [HSS].

Of the two general treatments of the movers' problem, the most important is the seminal paper by Schwartz and Sharir [SSh]. Unfortunately, the running time exponent of the Schwartz-Sharir algorithm is daunting for even simple problems. Their method partitions an n-dimensional space into simple pieces or cells, whose connectivity can easily be determined. Unfortunately, the tools they used, namely cell decompositions and Sylvester resultants, have poor complexity bounds for problems in high dimensions. As the number of dimensions of the space increases, both the number of cells, and the degree of the algebraic surfaces enclosing them, grow doubly exponentially. For motion planning, dimension translates into the number of degrees of freedom of the robot, which will typically be in the range three to six.

The roadmap algorithm described in this thesis solves the problem in exponential time. In place of cell decomposition, the roadmap algorithm uses a coarse *stratification*, and instead of Sylvester resultants, it uses *multivariate* resultants. Both these tools are new to computer science, but are straighforward to understand and apply. The drop from double to single exponential growth is crucial, since it implies that planning problems with just a few degrees of freedom can be solved in low polynomial time as a function of the environment complexity. For example, if we plan for the first three links of a typical Puma-type industrial robot, we get an $O(n^3 \log n)$ algorithm, where n measures the number of obstacle vertices, edges and faces.

Some of the mathematics will be new to most readers, and some familiarity with differential topology or geometry is very valuable. But the book is designed to be self-contained, and important notions (like transversality) are described with the aid of figures wherever possible. Those with some knowledge of the relevant mathematics will find that we do not need any deep results. Rather we have taken a collection of basic theorems and lemmas and reversed the usual abstraction process to bring them back to a constructive form.

We borrow heavily from the theory of singularities of maps between manifolds, in particular the results on stratified sets. These seem to be a very useful tool for computational geometry, and are a natural generalization of planar arrangements. From early results in elimination theory, we resurrect the multivariate resultant as the basic computational tool for all the algebraic algorithms in the thesis. Although a century old, the utility and efficient constructibility of the multivariate resultant have not been recognised previously. Results from these two areas are confined to a single chapter each, resultants being used in chapter 3, and singularity theory in chapter 4. With the exception of chapters 1 and 2, which give general background, each chapter can be read independently.

1.1 Robot Motion Planning Problems

1.1.1 The Generalized Movers' Problem

We saw above that the most basic motion planning problem is the generalized movers' problem (also called the find-path problem, obstacle avoidance problem) where the goal is to find any collision-free path. It is really a class of problems that includes motion planning for arbitrary robot arms or collections of arms, as well as objects (e.g. spacecraft) moving freely in space, and some types of mobile robot moving in two dimensions. A free object has six degrees of motion freedom, three translational and three rotational. A robot arm on the other hand, has several links which can move, but their motion is highly constrained by the joints. The joint between two links normally allows the links only one degree of relative motion freedom. This makes the free polyhedron and the robot arm seem like very different systems, but from a certain perspective, they are identical. This perspective is *configuration space.*

In order to describe the configuration of the polyhedron, we attach a reference frame (a set of local coordinate axes) to it. Then we can represent the configuration of the free polyhedron with six numbers, say the x, y, and z coordinates of the origin of the object's reference frame, and three angles which specify the orientation of the frame. All these numbers are measured with respect to some other reference frame fixed in the environment. These six numbers also define a point in a six-dimensional space called the configuration space of the polyhedron. Points in configuration space represent configurations of the polyhedron, and a path in configuration space represents a continuous motion of the object.

In contrast, we cannot feely place the links of the robot arm anywhere in space because of the constraints imposed by the links. However, if the links are not connected in a loop, we can freely choose all of the angles (or lengths for sliding joints) of the joints. Each choice of joint parameters completely determines the position and orientation of all of the links in the arm. Instead of the true coordinates of all the links, the configuration space now has generalized coordinates, which are the joint angles. If the robot has n variable joint parameters, then its configuration space is n-dimensional. Once again, a continuous motion of the robot corresponds to a path through configuration space.

If there are obstacles near the robot (or free object), then these impose additional constraints on motion. In configuration space they define forbidden regions, which are the configurations where some part of the robot overlaps an obstacle. The set of configurations where the robot is clear of obstacles is often called its *free-space*. An example of a two-link planar robot moving among obstacles, and its free-space is given in figure 1.1. To plan an obstacle-avoiding motion in physical space, we must find a path through configuration space which lies wholly within free space. Once the obstacles are mapped to forbidden regions in configuration space then, motion planning for both the free object and for the robot arm are the same process. The explicit use of configuration space for planning collision-free motions was first proposed by Lozano-Pérez in [Loz].

In virtually all cases the constraints on motion have an algebraic description, and the free-space can be defined as a "semi-algebraic set" [SSh], [Ca86] (Semi-algebraic sets will be defined and studied in the second part of this introduction). In principle, the set of free configurations can then be broken down into simple pieces or "cells" using a method called cylindrical algebraic decomposition [Col], [BKR]. By determining which cells are adjacent using the methods of [SSh], [KY], or [Pri], the existence of a collision-free path between two configurations can be determined. That is, if there is a chain of adjacent cells between the cell containing the start configuration and the cell containing the final configuration, then a collision-free

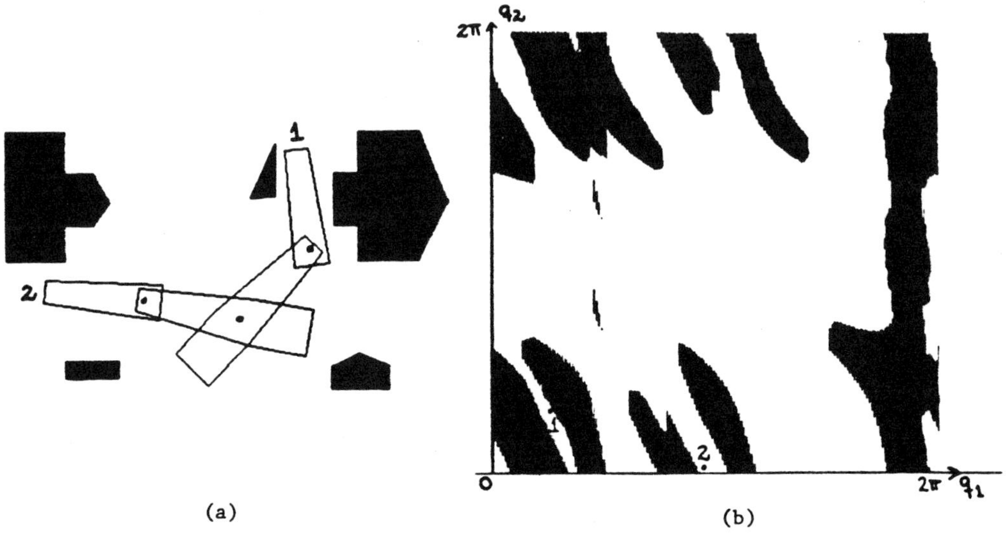

Figure 1.1: (a)[3]A two link planar manipulator, and (b) the constraints on motion in its configuration space.

path exists.

Unfortunately, these cell decomposition algorithms are also capable of solving a provably hard problem, deciding the theory of the real numbers. The decision problem for this theory is hard for exponential time with a polynomial number of alternations (a result of Berman's [Ber]). This is strong evidence that these algorithms will require double exponential time in the worst case. The average case performance of cell decomposition algorithms is not much better than their worst case, so the double exponential bound is a realistic estimate of running time.

The crucial input parameter in determining the running time is number of variables, or the dimension of the configuration space. If this number is fixed, then the running time is polynomial in the other measures of the size

[3]This figure courtesy of Prof. T. Lozano-Pérez

of the problem (such as the number of obstacles). However, this is disguising the fact that the exponent of the polynomial is very large, and itself grows exponentially with the number of variables.

Reif [Rei] first showed that the generalized movers' problem is PSPACE-hard, which suggests that in the worst case, single exponential time may be required by *any* algorithm which solves the problem. Other authors have since produced PSPACE hardness results with many movable objects in the plane [HSS]. Both these results use constructions with many degrees of freedom, and this number grows with the size of the problem.

On the other hand, there do exist many efficient algorithms for restricted classes of motion planning problems [SY], [SHS]. These algorithms work for systems with a fixed number of degrees of freedom, or for certain very symmetric problems (like coordinated motion of many disks) with an arbitrary number of degrees of freedom. In the latter case, they have running time exponents which are *linear* in the number of variables. This suggests that it may be possible to solve the general problem in exponential time, with exponent linear in number of degrees of freedom. This is quite useful in practice, where the number of degrees of freedom is best thought of as a small constant rather than a input parameter. We present such an algorithm here. In fact the running time of our algorithm equals or improves the worst-case performance of many of the special purpose algorithms referred to above, especially with regard to algebraic complexity. From Chapter 5, the running time of the algorithm is

$$T_{R_1} = \left(n^r(\log n(2d)^{O(r)} \log^3 w + (2dr)^{O(r^2)} \log^2 w)\right) \qquad (1.1)$$

where r is the number of degrees of freedom, n is the number of constraints (product of the number of edges, vertices etc. in the obstacles and the moving object), and d and w are respectively the degree and coefficient magnitude of the constraint polynomials.

The bound may be weakened to a product of "geometric" part $O(n^r \log n)$, measuring growth with environment complexity, and an "algebraic" part $((2dr)^{O(r^2)} \log^3 w)$ measuring growth with degree and coefficient size. Previous algorithms for the general problem exhibit double exponential growth in both parts. There are two new tools which make single exponential bounds possible. The first of this is the notion of stratification, which is a partition of S into smooth pieces (manifolds). We use the coarsest possible stratification to limit the geometric complexity of the algorithm. The second tool is the multivariate resultant, which allows simultaneous elimination of variables in single exponential sequential time, or polynomial parallel time. We also derive from bounds on the size of the resultant an improved gap theorem. This theorem is used throughout the roadmap algorithm to reduce computation with algebraic numbers to computation with finite precision binary numbers.

1.1.2 The Shortest Path Problem

The shortest path problem is of interest in robotics because it is the simplest instance of a minimum cost path planning problem. The *L^p-shortest path problem*, may be defined in two or three dimensions as the problem of finding the shortest obstacle-avoiding path between two given points under an L^p metric, with polygonal or polyhedral obstacles respectively. The L^p norm of a vector $v = (x, y, z)$ is denoted $||v||_p$, and is defined by

$$||v||_p \;=\; \sqrt[p]{|x|^p + |y|^p + |z|^p} \tag{1.2}$$

and the L^p distance between vectors u and v is $||u - v||_p$. The usual euclidean norm is $||\cdot||_2$. Efficient algorithms for two-dimensional euclidean shortest path have been known for some time. This problem was first studied by Lozano-Pérez and Wesley [LW]. No upper bounds were given in their paper, but their algorithm should run in $O(n^3)$ time. Improvements were given by Sharir and Schorr [SSc], and by Reif and Storer [RS] who gave an $O(n(k + \log n))$ time algorithm, where k is the number of convex parts of the obstacles.

While the euclidean shortest path has efficient solutions in two dimensions, the three-dimensional problem is much more difficult. There have been a number of papers on restricted versions of the problem. Mount [Mou], and Sharir and Baltsam [SB] gave polynomial time algorithms for environments consisting of a small fixed number of convex polyhedra. Papadimitriou [Pap] gave a polynomial time approximate algorithm which finds a path at most a small multiplicative constant longer than the shortest path. The general problem has been dealt with by Sharir and Schorr [SSc] who gave an $2^{2^{O(n)}}$ algorithm by reducing the problem to an algebraic decision problem in the theory of real closed fields. The best bound is due to Reif and Storer [RSt] who gave an $2^{n^{O(1)}}$-time, $(n^{O(\log n)})$-space algorithm by using the same theory but with a more efficient reduction. To date, in spite of the absence of efficient algorithms for the general problem, no lower bounds were known.

In chapter 6, we show that the single source multiple-destination shortest path problem may lead to exponentially many distinct shortest path classes, and that the single-source single-destination problem is NP-hard. Our results hold for any L^p metric.

1.1.3 Dynamic Motion Planning with Velocity Bounds

We consider motion planning for a robot with a fixed number of degrees of freedom in an environment in which the obstacles are moving, and in which the robot has constraints on its velocity. If there are no motion constraints, and the trajectories of obstacles can be described algebraically, then the problem is solvable in polynomial time [RSh]. However, if velocity limits are added, the problem is more difficult. In [RSh] it was shown that motion planning in 3-d with rotating dynamic obstacles is PSPACE-hard in the presence of velocity bounds, and NP-hard in the absence of bounds. However, the rotating obstacles have non-algebraic trajectories. Here we give the much stronger result that motion planning for a point in the plane with velocity bounds is NP-hard, even when the moving obstacles are convex polygons moving with constant linear velocity without rotation.

We define the 2-*d asteroid avoidance problem* as the problem of determining a collision-free path for a point in the plane with bounded velocity magnitude, with convex polygonal obstacles moving with fixed linear velocity (no rotation). The obstacles are assumed not to collide.

Using a proof technique similar to that for the shortest path problem, we show that the 2-d asteroid avoidance problem is NP-hard, and that at certain times, the set of reachable positions in the plane may consist of exponentially many connected components.

1.1.4 Compliant Motion Planning with Uncertainty

We also treat motion planning with uncertainty, where the objective is to produce a plan which is guaranteed to succeed even if the robot cannot perfectly execute it due to control error. With control uncertainty, it is impossible to perform assembly tasks which involve sliding motions using only position control. Robot control uncertainty is significant and has traditionally biased robot applications toward low-precision tasks such as welding and spray-painting. To successfully plan high-precision tasks such as assembly operations, uncertainty must be taken into account, and other types of control must be used which allow *compliant motion.* Compliant motion ([Ino], [Mas]) occurs when a robot is commanded to move into an obstacle, but rather than stubbornly obeying its motion command, it complies to the geometry of the obstacle.

Compliant motion is possible only with certain dynamic models. Two common models are the generalized spring and generalized damper models, [Buc87], [Mas], [Erd]. Our proof will succeed with either of these models, but the one we will use is a simplified version of the damper model described in [LMT]. We assume that our environment describes the configuration space of the robot, so that the robot itself is always a point. The planned path consists of straight-line commanded motions each for a fixed time interval. That is, at the ith step, the point is commanded to move at velocity v_i for time t_i. Because of control uncertainty however, the point actually moves with a velocity v_i^{free} which lies in a ball of radius ϵv_i about the commanded velocity, i.e.

$$||v_i^{free} - v_i|| < \epsilon v_i \tag{1.3}$$

Without loss of generality, we assume all v_i have unit magnitude, and scale the t_i accordingly.

For a compliant motion, the object moves along an obstacle surface with a sliding velocity v_i^{slide} which is the projection onto the surface of some v_i^{free} satisfying (1.3). This v_i^{free} must point into the surface, as shown in figure 1.2. Figure 1.2 (a) shows a side view of a peg-in-hole insertion, while 1.2(b) shows the obstacle in configuration space seen by the reference point on the peg. The motion of the peg (without rotation) can be determined from the motion of the reference point based on generalized damper dynamics. We will not consider further details of the dynamic model since they are not

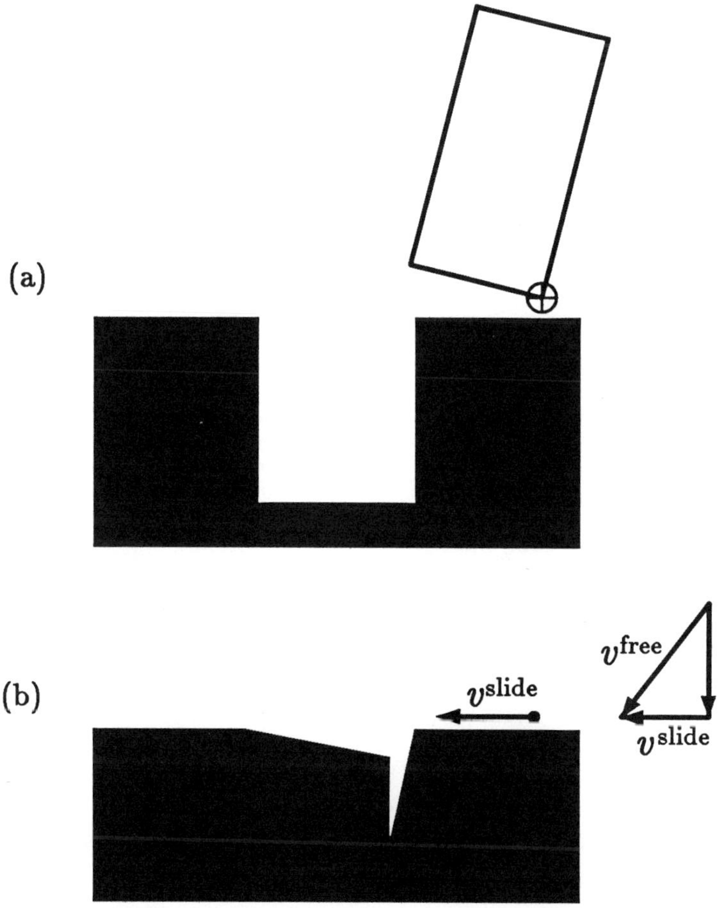

Figure 1.2: (a) Peg and hole environment, (b) Configuration space showing locus of reference point of peg during compliant motion.

necessary for our proof, but we recommend the reference [LMT].

The *Compliant Motion Planning with Uncertainty* problem is to find a sequence of motions v_i, which is guaranteed to move every point in a polyhedral start region S into a polyhedral goal region G, moving among and possibly sliding on polyhedral obstacles. Each motion is subject to a velocity uncertainty ϵ as described above. We consider a polyhedral start region because there will inevitably be some uncertainty in the initial position of the point. That is, we know that p is initially somewhere in S but we cannot know where exactly. However, because our plan succesfully moves every point in S into G, it is guaranteed to move p from its actual position into G. In [Nat] it is shown that this problem is PSPACE-hard. Using some of the new techniques in this paper, we are able to show that 3-d compliant motion planning is non-deterministic exponential time hard. We believe this to be the first instance of a *provably* intractable problem in robotics.

1.2 Algebraic Decision Procedures

Algebra and Geometry are intimately linked. While our interest is in geometric problems that arise in robotics, our descriptions of that geometry are algebraic. Sooner or later we must answer algebraic questions involving systems of equations, either numerically or symbolically. Here we examine several existing decision methods, and explore their suitability, in fairly general sense, for dealing with the kinds of algebraic problem that arise in robotics. We concentrate on those techniques whose performance is well-characterized theoretically, or which have been extensively studied experimentally. We consider theories of the reals and complexes, as well as algorithms that deal directly with polynomial ideals.

By the "theory of the reals", we mean the first order theory of real closed fields with $+$, $*$, and $>$. Formulae in the language are built from boolean combinations of polynomial inequalities. Both existential and universal quantification over the reals are allowed, making this a very powerful language. A typical formula in this language is:

$$\exists x \forall y (y^2 \geq x) \wedge (-y^2 \leq x) \tag{1.4}$$

A formula with n free variables defines a *semi-algebraic set*, in $\Re^n$, which is the set of variable values for which the formula is true. Tarski [Tar] showed that every semi-algebraic set defined by such a formula also has a defining formula which is free of quantifiers. Thus we can say that the collection of semi-algebraic sets is closed under finite union, intersection, complement, and elimination of quantifiers. Semi-algebraic sets seem to be a natural choice for real geometry and robotics, and it is easy to represent polyhedra, robot kinematic constraints, and CSG models as semi-algebraic sets.

We will also be considering the theory of the complex numbers, where there is no notion of ordering. The theory of the complexes lacks a $>$ operator, and formulae with free variables define instead *constructible sets*. Constructible sets are closed under finite union, intersection, complement and quantifier elimination. They are somewhat less powerful for robotics and real geometry, since there is no way to define the interior or exterior of an object. Constructible sets over the complexes can also be represented as semi-algebraic sets over the reals, using two real variables to represent one complex number.

Finally, we define an *algebraic set* as the set of common zeros of a system of polynomials. Algebraic sets are considerably more restrictive than semi-algebraic or constructible sets, and are closed only under finite union and intersection. They are not in general closed under negation or quantifier elimination, although if the polynomials defining an algebraic set are homogeneous in some of the variables, simultaneous elimination of those variables defines an algebraic set. Algebraic sets are constructible. We can rank these representations in increasing order of expressive power:

$$\text{Algebraic Sets} \subset \text{Constructible Sets} \subset \text{Semi-Algebraic Sets}$$

1.2.1 The Theory of the Reals

As might be expected, the decision problem for the theory of the reals is difficult, and Berman [Ber] has shown that it is hard for exponential time with a polynomial number of alternations. By Tarski's [Tar] famous result, it is decidable, although the procedure originally given by Tarski is non-elementary. Since then there have been a number of improvements, first by Seidenberg [Sei], and later by Cohen [Coh], and eventually, the double-exponential time method of Collins [Col]. The crucial parameter here is the number of quantifiers. In the light of Berman's lower bound, this upper bound seem to be the best possible. There have been a number of improvements on Collins procedure, although its basic sequential time bound has not been improved. An improvement on the space required to decide the language was given by Ben-Or, Kozen and Reif, [BKR], who showed that the theory could be decided in single-exponential space (and therefore in single-exponential parallel time). Also worthy of note is Grigoryev's recent result [Gri], which gives a decision procedure which is doubly-exponential only in the number of quantifier *alternations*.

The theory of the reals has been a powerful tool in establishing upper bounds for new problems in robotics, although there seem to be no current implementation of any of the above decision algorithms which is fast enough to constitute a practical solution to these problems. For example, the existence of an obstacle-avoiding path of a certain length in three dimensions can be formulated as a decision problem in the theory of the reals. It is known that if such a path exists, then there exists a path of that length which consists of linear number (in the environment size) of straight line segments. Given a predicate which indicates if a point is in free space (see chapter 2), we can construct a similar predicate for line segments using universal quantification over a variable which parametrizes position along the segment. From this predicate, we can construct a similar predicate for a piecewise straight line path by "guessing" a polynomial number of via points using existential quantification. We now have a decision problem for the theory of the reals, and with a little work, we can actually find the shortest path. Since the formula has polynomial size, and only one quantifier alternation, by Grigoryev's result, we obtain an immediate single-exponential time upper bound for the shortest path problem. A similar technique yields a single-exponential time upper bound for the problem of dynamic obstacle avoidance with velocity constraints. Given the new lower bounds presented in this thesis (chapter 6), these upper bounds would seem to be best possible.

In addition, for a formula in n variables (quantifiers), Collins' technique actually builds a decomposition of the space $\Re^n$ of variable values. This decomposition consists of very simple geometric pieces called cells. If cell adjacency information is added to Collins algorithm, one obtains substantial topological information about the region in quantifier space in which the defining predicate (i.e. the formula with quantifiers removed, which is just a predicate in n variables) is true. This predicate could be, for example, the set of interference-free configurations of a robot manipulator. Such a cell adjacency procedure was given by Schwartz and Sharir [SS] in their general motion planning algorithm, and an improvement (to exponential space) was given by Kozen and Yap [KY]. However, these methods have double-exponential growth in time with number of variables, which is substantially worse than the PSPACE lower bound of Reif [Rei]. It seems that Collins method is perhaps overkill for this problem, and we should seek a less powerful but more efficient tool.

In very simple terms, virtually all the decision algorithms which are capable of deciding the theory of the reals (with the significant exception of Grigoryev's method [Gri], see below) operate by successive elimination of variables. Each such elimination step leads to at least quadratic growth in the number of polynomials, hence the double exponential upper bounds. By contrast, the decision procedures for the existential theory described next perform simultaneous elimination of several variables, with dramatic improvements in runtime.

1.2.2 Existential Theories, Complex Numbers and the Reals

Before describing these methods we should mention the work of Lazard [Laz81], who gave a single exponential time algorithm for solving systems of polynomial equations. While not a decision procedure per se, his algorithm was the first to solve an algebraic problem in several variables in less than double exponential time in the number of variables. It is also an important subalgorithm in the decision procedures of Chistov and Grigoryev [CG] and Grigoryev and Vorobjov [GV]. A recent algorithm by Renegar [Ren] also gives a single-exponential bound for solving systems of equations, and this algorithm incorporates more realistic assumptions about computation with real numbers on a computer than did Lazard's.

Taking the existential theory of the complex numbers first, it differs from the theory of the reals in that it has only existential quantification. Furthermore, since the complex numbers lack an ordering, the only type of comparison is equality, and only constructible sets can be defined by formulae with free variables. It is considerably less "natural" for geometry than theories with real quantification, since quantification normally occurs over physical dimensions or positions. The lack of a > operator is also a significant handicap. Deciding the theory can be easily shown to be NP-hard (introduce polynomials $x_i^2 - x_i = 0$ so that all variables are 0-1 valued, then it is easy to encode a SAT formula with $\times$ representing AND, and $+$ representing OR). In spite of its limitations, this theory seems to have some interesting applications, it includes all the theorems in Wu's geometry [Wu] for example. In [CG], Chistov and Grigoryev gave a single exponential time decision procedure for this theory, the first sub-doubly exponential time procedure for an algebraic theory.

Extending this work, Grigoryev and Vorobjov [GV] gave a single exponential time decision procedure for the existential theory of the reals. This theory lacks only universal quantification compared to the full theory of the reals, and is strictly more powerful than the existential theory of complex numbers.

The key element of both these methods which allows single exponential time bounds is the simultaneous elimination of several quantifiers using a variation of Lazard's method. Simultaneous elimination of variables can also be used in the full theory of the complexes or reals to eliminate contiguous sequences of quantifiers of the same type [CG], [Gri]. Chistov and Grigoryev also used generic points to represent irreducible components of algebraic sets. Hong [Hon] later gave a method for computing generic points as binary numbers using a gap theorem (see chapter 4).

Finally, we should mention here Wu's method [Wu], and Hong's method [Hon]. Both have been used as tools in the proofs of some elementary geometry theorems. Both deal only with certain "generic" algebraic sets, and the number of polynomials is restricted, boolean connectives are not available, and quantification occurs over the complexes. These methods assume as input a certain triangular form for the input polynomials, which must be carefully hand crafted for input to the algorithms. It is not clear how difficult it is to construct this triangular form from an arbitrary system of polynomials, and this limits the applicability of these methods for robotics, where more general polynomials arise.

Both methods seek to determine whether a consequent polynomial vanishes at all the common zeros of a system of hypothesis polynomials. They may be thought of as restricted existential theories, (since they test for the existence of a point at which the hypothesis polynomials vanish but the consequent does not) which operate on algebraic sets. Wu's method tests the consequent polynomial by pseudo-dividing with the hypothesis polynomials, requiring human intervention to check irreducibility of certain intermediate polynomials. Hong's method evaluates the consequent polynomial at *generic points* of the irreducible components of the variety defined by the hypothesis polynomials. Although complexity analysis of Wu's and Hong's methods have not been given, both appear to be single exponential. The theory that they are capable of deciding is NP-hard, since the hardness proof given above for the existential theory of the complexes still holds here.

1.2.3 Standard or Gröbner Bases

So far in our discussion we have been considering algorithms for manipulation of geometric objects described by polynomial equalities or inequalities. We have treated systems of polynomials simply as a convenient representation of geometric objects called algebraic or semi-algebraic sets. However, systems of polynomials also define algebraic objects called ideals, which have additional structure. There is a many-to-one correspondence between ideals and algebraic sets, and a one-to-one correspondence between algebraic sets and a particular subclass of ideals, called self-radical ideals, which is given by Hilbert's Nullstellensatz (see, for example [Wae] or [Mum]).

Gröbner or standard basis algorithms allow manipulation of polynomial ideals directly, and allow membership, intersection, union, and division operations on arbitrary ideals. They can also be used to find the solutions of systems of polynomial equations. Given a basis for an ideal i.e. a set of polynomials which generate the ideal, a standard basis is a set of polynomials which generates the same ideal, and which allows an easy reduction or "division" operation on other polynomials. All the other algorithms we have described in this section compute properties of algebraic sets only, and ignore the algebraic structure of the ideals. This distinction has important computational implications, and it seems that decision problems involving the algebraic structure of the ideals are intrinsically harder than problems involving only the coarser geometric structure. For example, Mayr and Meyer [MM] have shown that the most basic problem in ideal theory, ideal membership, is exponential space hard. On the other hand, we have seen in this section single exponential time algorithms for numerous problems involving only the geometry of the algebraic sets.

Upper bounds for standard basis algorithms are rather crude at this stage, and are more often given as bounds on degree rather than on time for computation of the basis. The size (measured by degree) of a standard basis may be double-exponential in the worst case. This makes them less than attractive for application in robotics, where it is difficult to make guarantees about the form of the input polynomials. However, certain types of ideal seem to be easier to deal with than others, in particular, decision problems involving self-radical ideals seem to have lower complexity. While this result remains to be proved, a conjecture along these lines is given in [BS] based on the regularity of syzygies for the ideal membership problem. Thus it may still be possible to make use of standard basis algorithms in computational geometry by computing a basis for the radical of an ideal rather than the ideal itself. It may also be possible to apply the basis algorithms to perturbed or "generic" polynomials, since their average case performance is often much better than the worst case would suggest. Standard basis algorithms are also related to Lazard's method (and to the multivariate resultant), and this connection is explored in [Laz83]

The notion of standard ideal basis which we are using here is due to Hironaka [Hir]. The term Gröbner basis was introduced shortly afterward by Buchberger [Buc65], who also gave a computational procedure for their construction. Since that time there has been a considerable literature on the computation and application of standard bases, and this work now forms a active subbranch of computer algebra. A recent survey of the area is given in [Buc85].

1.3 Overview of the Thesis

Chapters 2 through 5 of this thesis deal with algorithms for robot motion planning, while chapter 6 gives lower bounds for some basic motion planning problems. We begin in chapter 2 by defining the motion constraints in configuration space, and then give algorithms for planning in configuration space in chapters 4 and 5. Chapter 3 is more or less self contained, and describes the algebraic techniques necessary for the roadmap algorithm of chapter 4.

The configuration space approach, [Udu], [Loz] allows motion planning for a complex solid object to be reduced to motion planning for a point in the object's configuration space. We first choose some parametrization of the object's configuration, and then the configurations at which the object overlaps obstacles define a region in its configuration space called the configuration obstacle. In chapter 2, we define the configuration obstacle for free polyhedra and for robots with polyhedral links. We also give a fast collision detection algorithm using this formulation. The configuration obstacle predicates defined in this chapter provide the "front end" to the planning algorithms described in later chapters. The particular form of these predicates leads directly to the definition of the simplified Voronoi diagram in chapter 5.

Chapter 3 is independent of the other chapters of the thesis, and develops a new algebraic tool called the multivariate resultant. The multivariate resultant is a generalization of the usual (Sylvester) resultant and the matrix determinant, and allows efficient simultaneous elimination of several variables from a system of polynomial equations. In this chapter we give a basic recurrence defining the resultant, and two methods for its computation. From the second of these methods, we derive an improved gap theorem. Gap theorems give lower bounds on the difference between distinct algebraic numbers defined by polynomial systems of a certain size. They allow "exact" computation with algebraic numbers by using finite precision binary approximations. We also give in this chapter a fast method for solution of systems of polynomial equations in several variables.

Chapter 4 contains the main result of the thesis, which is an algorithm for the generalized movers' problem which runs in single-exponential time, with exponent equal to the number of degrees of freedom of the robot. The algorithm computes a one-dimensional skeleton or roadmap of the set of free configurations. We first give a mathematical definition of the roadmap, using techniques from differential topology. In particular, we make use of the notion of stratified sets to give us a coarse partition of semi-algebraic sets into simple pieces (manifolds) of fixed dimension. The basic roadmap is built from sets of curves which are linked together at critical values in a manner reminiscent of Morse theory. We then refine this construction using "linking curves". These curves can be computed in linear time, and allow start and goal points to be rapidly joined to the roadmap.

Chapter 5 deals with heuristic techniques for the generalized movers' problem. We consider only provably good heuristics, that is, heuristics which may improve average running time without sacrificing correctness, accuracy, and if possible, worst case running time. We described a method for reduction of the dimension of the semi-algebraic set S defining free-space. We do this by defining a new kind of Voronoi diagram, called the simplified Voronoi diagram. This diagram differs from conventional Voronoi diagrams in that it uses the constraint predicates themselves as a measure of distance. This means that the algebraic complexity of the diagram is the same as for the boundary of free-space, and much lower than diagrams based on euclidean distance. We show in this chapter that the diagram is complete for motion planning, that is, if there exists a path between any two points on the diagram, then this path can be deformed so that it lies entirely on the diagram. Parts of this chapter represent joint work with Bruce Donald of MIT and Cornell University.

Chapter 6 complements the other chapters of the thesis by giving lower bounds on several natural extensions of the generalized movers' problem. All the results in this chapter involve systems with three degrees of freedom or fewer. The main result of this chapter is that the problem of finding the shortest path in a polyhedral environment (under any L^p metric) is NP-hard. We also show that motion planning with moving obstacles in two dimensions is NP-hard if there are limits on the robot's velocity. Finally we show that compliant motion planning with uncertainty is hard for non-deterministic exponential time. Parts of this chapter represent joint work with John Reif of Duke University.

Chapter 2

Motion Constraints

In this chapter we compute the predicates that define configuration space obstacles. We consider only the case where object and obstacles are polyhedral. The configuration space obstacle for a pair of polyhedra is simply the set of configurations at which the two polyhedra overlap. The predicates we are looking for are called *overlap predicates*, and are true at exactly those configurations where there is overlap. We consider both the configuration spaces of free polyhedra (polyhedra with translational and rotational degrees of freedom), and multi-link robot arms with polyhedral links. The overlap predicates are computed in two steps, and the first step is independent of the configuration space. First we define static overlap predicates for a pair of polyhedra, (which could be either either free or part of a linked robot) assuming that they share a common coordinate frame . The static predicates are functions of the *feature coordinates* of the polyhedra, such as vertex coordinates, face normals etc. The overlap predicates are obtained by substituting for the feature coordinates of one (or both) polyhedra as functions of the configuration parameters. Later in the chapter we use the overlap predicates in an efficient collision detection algorithm.

The static overlap predicates for two polyhedra are constructed from *primitive predicates* using logical connectives. A primitive predicate involves only a single pair of features, one from each of the two polyhedra, such as a vertex of one polyhedron and a face of the other. We will derive two different forms for the static overlap predicate which have different logical structure.

One of the predicates is defined only for convex polyhedra. While it would be possible to select one of these as being more efficient or compact, they exhibit very different behaviour in practice when used with tree-pruning algorithms, and so both are included here. Both forms are constructed from the same set of primitive predicates. Each primitive predicate is a low degree polynomial inequality.

Once we have a static overlap predicate (in one of the two forms mentioned above), we substitute for all of the feature coordinates as functions of the configuration space parameters, giving the final overlap predicates. Since the static overlap predicates are logical combinations of primitive predicates, substitution only affects the primitive predicates, not the logical structure of the overlap predicate. Since the primitive predicates are low-degree polynomial inequalities, and feature transformations are (low-degree) polynomial functions of configuration parameters, the final overlap predicates are also logical combinations of low-degree polynomial inequalities.

The advantage of this two-step process is that the static overlap predicates need only be defined once, and are independent of the configuration space of the system. We need only concern ourselves with the parametrization of configuration space when we are computing the feature transformations. While we treat only two basic (but important) cases in this chapter, it is easy to generalize the method to an arbitrary configuration space. Finally, there are only three types of transformation that need to be computed, one each for vertices, edges and faces.

In section 2.1 we define the two types of static overlap predicate for polyhedra. The forms of static overlap predicate are called conjunctive and disjunctive respectively, according to the type of combination that is applied at the root of the predicate formula. Both types are defined for convex polyhedra, but only the disjunctive form generalizes to the non-convex case. Although the two forms test for the same condition (for convex polyhedra, at least), and have roughly the same number of primitive predicates, the conjunctive form leads to deeper predicate trees, and these seem to prune more readily in practical examples. In both cases, the size of the overlap predicate is the product of the number of features on one polyhedron with the number of features on the other.

In section 2.2 we show that the primitive predicates correspond to the basic types of contact between polyhedra. This section makes clear the relationship of the method described here with the constraint representation used by Lozano-Pérez [Loz]. In the latter case, contact conditions are derived first and used to define the boundaries of the configuration space obstacle. Auxiliary conditions must be added which test whether contact is possible at a given orientation. The location of a point in the interior/exterior of the obstacle is not given by a predicate. In our case, the configuration obstacle is obtained by parametrizing a static overlap predicate, so that interior/exterior information is immediate, and contact conditions derive naturally from the boundary of the region defined by this predicate.

In section 2.3, we give explicit forms for the overlap predicates for a free polyhedron. We use a quaternion representation for the orientation of the polyhedron which gives us constraints with low algebraic degree. We briefly review some basic properties of quaternions, and introduce the "four-cubes" representation which is a topologically faithful three-dimensional representation for the orientation of the object. We then give expressions for transformed object features, and for the primitive predicates after substitution of these features.

In section 2.4 we give a fast and simple algorithm for collision detection. The algorithm accepts an algebraically-defined path in configuration space and computes all the collision points of the object with obstacles as it follows this path. By substituting a parametrized path into the overlap predicate, we obtain a predicate which is a function of a single parameter, which is the distance along the path. To compute collision points, we need only find the true-false transitions of this predicate. The algorithm runs in pseudo-linear time in the size of the predicate formula, which for the disjunctive form is proportional to the product of the number of object features and the number of obstacle features, where object and obstacles are arbitrary polyehdra. We extend the single moving object algorithm to several moving objects moving along known paths in section 2.5.

Finally, in section 2.6 we derive the configuration obstacle for a multi-link robot arm. Here the configuration space is parametrized by the joint angles (or displacements) of all the robot's joints. For rotary joints, we use an algebraic parametrization of the joint angle so that the overlap predicate will be algebraic. We use products of homogeneous matrices based on Denavit-Hartenberg descriptions of the robot [DH] to compute the relative transformation between links of the manipulator.

2.1 Object-Obstacle Constraints

In this section we give two predicates for overlap of a polyhedral object with polyhedral obstacles. For the conjunctive predicate, object and obstacles are assumed to be convex. Both the conjunctive and disjunctive form are AND-OR trees with roughly the same number of leaves in the convex case. In the non-convex case, it is still possible to use the conjunctive form by decomposing the polyhedra into convex pieces, but this can lead to quadratic blow-up in the number of features [Cha]. The *disjunctive form* has an OR-node at its root. The *conjunctive form*, which is actually derived from a non-overlap test, has an AND-node at its root. In the algorithms described in this thesis, we have found the conjunctive form to be the most useful, because it generates deeper AND-OR trees which can be more extensively pruned. However, since the disjunctive form works for arbitrary polyhedra without any increase in complexity, it should be preferable in some situations.

2.1.1 Disjunctive Form

Lemma 2.1.1 *Let A and B be two arbitrary (connected) polyhedra, and let $\mathbf{p}_A$ be any vertex of A, and $\mathbf{p}_B$ be any vertex of B. Then A and B overlap if and only if either* (i) *$\mathbf{p}_A$ is inside B, or* (ii) *$\mathbf{p}_B$ is inside A, or* (iii) *an edge of A pierces a face of B, or* (iv) *an edge of B pierces a face of A.*

Proof Clearly if any of these conditions is true, then A and B overlap. Conversely, if A and B intersect, then either the intersection equals A, i.e. $A \subset B$, in which case (i) holds, or $B \subset A$, in which case (ii) holds, or the intersection equals neither A nor B. In the third case, the boundaries of the two polyhedra intersect. This implies that some face of A intersects a face of B. This intersection is bounded by either an edge of A, in which case this edge and the face of B intersect, so that (iii) is true or the intersection is bounded by an edge of B, in which case (iv) is true. □

If the polyhedra are not connected, then we need one point p_A within each connected component of A, and one point p_B in each connected component of B, for the above test to succeed.
For the derivation of the predicates, we need the following definitions:

Definition. For any face f of a polyhedron A, the *affine hull* $\bar{f}$ of f is the plane which contains f.

The affine hull of a face f defines two closed half-spaces, one of these is on the interior side of f, and is called the *interior half-space* of f, and denoted $\bar{f}^-$.

From the above lemma we can define a predicate for overlap of two arbitrary polyhedra A and B as

$$O_{A,B} = O_{p_A,B} \vee O_{p_B,A} \vee \Bigg(\bigvee_{\substack{e_i \in \\ \text{edges}(A) \\ f_j \in \\ \text{faces}(B)}} O_{e_i,f_j} \Bigg) \vee \Bigg(\bigvee_{\substack{e_j \in \\ \text{edges}(B) \\ f_i \in \\ \text{faces}(A)}} O_{e_j,f_i} \Bigg) \tag{2.1}$$

where $O_{\mathbf{p},A}$ is the predicate for overlap of the point $\mathbf{p}$ and the polyhedron A, and $O_{\mathbf{e},f}$ is the overlap predicate for an edge e and a face f. If A is convex, the point-polyhedron predicates have a simple conjunctive form:

$$O_{\mathbf{p},A} = \bigwedge_{\substack{f_i \in \\ \text{faces}(A)}} \neg A_{\mathbf{p},f_j} \tag{2.2}$$

where $A_{\mathbf{p},f}$ indicates *non*-overlap of a vertex $\mathbf{p}$ of B and the interior half-space of a face f of A and is given by

$$A_{\mathbf{p},f} = (\mathbf{n}_f \cdot \mathbf{p} - c_f > 0) \tag{2.3}$$

where $\mathbf{n}_f$ is the outward normal of f, and c_f is its distance from the origin. The non-convex case must be handled differently, and we postpone its derivation for the moment, and proceed to the edge-face test.

For the edge-face test, we assume that the face f is convex. If f is non-convex, then it can be decomposed into a union of a linear number of convex pieces in pseudo-linear time. Alternatively, we could make use of a CSG predicate for the face as defined in [DGHS], which has linear size. The point-in-polygon CSG predicate derived in [DGHS] can be used to test whether the intersection point of the edge with the plane of the face f lies within the face or not. To do this, the half plane tests in [DGHS] would be replaced by the left-right (C^+ and C^-) tests for e against the edges of f, which we describe below. Just as in equation (2.4) below, each such test would have to be split into two subtests, corresponding to the inward or outward orientation of e.

If the all faces f of the two objects are convex, we can use a simpler method than [DGHS]. In the following, we assume that faces have been decomposed into convex pieces, so that every face f is convex. In order for an edge e to pierce face f, one vertex on the edge must lie above the plane of the face and one must lie below. We must consider these two cases separately so we set

$$O_{e,f} = O^{out}_{e,f} \vee O^{in}_{e,f} \tag{2.4}$$

where $O^{out}_{e,f}$ indicates overlap of e and f with e pointing outward (positive inner product between e and the outward normal to f), and $O^{in}_{e,f}$ indicates overlap with e pointing inwards. We assume first that the head of e is above f and its tail is below. This condition can be easily tested since it derives from the predicates (2.3). Then we must decide whether the intersection point of the edge and the plane of the face f lies inside the face or not. If f is convex, then the intersection point is in f if and only if it lies inside the interior half-plane of every edge in the boundary of the face f. So the predicate

$$O^{out}_{e,f} = \neg A_{T,f} \wedge A_{H,f} \wedge \bigwedge_{e_i \in edges(f)} C^+_{e,e_i} \tag{2.5}$$

indicates overlap of e and f with e pointing outward, while

$$O^{in}_{e,f} = \neg A_{H,f} \wedge A_{T,f} \wedge \bigwedge_{e_i \in edges(f)} C^-_{e,e_i} \tag{2.6}$$

indicates overlap with e pointing into f, and where C^+ and C^- indicate which side of an edge e the intersection point lies. To derive them we need the following definitions: Let $\mathbf{d}_A$ and $\mathbf{d}_B$ be the vector directions of e_A and e_B respectively, and let $\mathbf{p}_A$ be any point on e_A, for definiteness, say the head of e_A. Let and H and T be the head and tail vertex respectively of e_B, then we define

$$\begin{aligned} C^+_{e_A,e_B} &= (H - \mathbf{p}_A) \cdot (\mathbf{d}_A \times \mathbf{d}_B) > 0 \\ C^-_{e_A,e_B} &= (H - \mathbf{p}_A) \cdot (\mathbf{d}_A \times \mathbf{d}_B) < 0 \end{aligned} \tag{2.7}$$

The vector triple products indicate whether the vectors $(H - \mathbf{p}_A)$, $\mathbf{d}_A$ and $\mathbf{d}_B$ form a right-handed coordinate system. To see that they give the information we want about intersection point position, notice that the value of $(H - \mathbf{p}_A) \cdot (\mathbf{d}_A \times \mathbf{d}_B)$ is not changed if we add any multiple of $\mathbf{d}_A$ or $\mathbf{d}_B$ to $(H - \mathbf{p}_A)$, since the cross product is orthogonal to both edge directions. This implies that while $(H - \mathbf{p}_A) \cdot (\mathbf{d}_A \times \mathbf{d}_B)$ is the triple product of the two edge directions with the difference vector between a *particular* point on the edge e_A and a *particular* point on e_B, the triple product is the same for *any* point on e_A and *any* point on e_B.

Suppose f is a face of A, and e_A is an edge bounding f. Then we can replace the points H and $\mathbf{p}_A$ in $(H - \mathbf{p}_A) \cdot (\mathbf{d}_A \times \mathbf{d}_B)$ respectively with the intersection point $\mathbf{p}_f$ of e_B with f, and the closest point $\mathbf{q}_e$ on the edge e_A to $\mathbf{p}_f$. This does not change the triple product. We can rearrange the triple product to the equivalent form $\mathbf{d}_B \cdot ((\mathbf{p}_f - \mathbf{q}_e) \times \mathbf{d}_A)$. Now the cross product here gives a vector which is *normal* to the face f, and points outward if and only if $\mathbf{p}_f$ lies to the right of the edge e_A, (left and right are determined here by viewing f from its exterior). Since we have already distinguished the cases $\mathbf{d}_B$ pointing into or out of the face f, by the dot-product of $\mathbf{d}_B$ with the outward normal of f via type-A predicates, the sign of the triple product tells us unambiguously whether the intersection point $\mathbf{p}_f$ lies to the left or right of the edge e_A. Assuming the edges are oriented in a counter-clockwise loop around the face f, the left or right information tells us whether the intersection point lies in the interior half-plane of the edge.

Returning to the test for containment of a vertex in a polyhedron, we make use of the edge-face test just described. Let $\mathbf{p}_B$ in B be the point we want to test, and create a new edge (really a ray) e_B with head $\mathbf{p}_B$, no tail, and direction d_B, where d_B can be any fixed "generic" direction (see below for justification for this). It is best if the direction of the vector d_B is fixed relative to A rather than B, so that it can be chosen not in the plane of any face of A, which eliminates some degenerate conditions.

If $\mathbf{p}_B$ is contained in A, then the ray e_B must intersect an odd number of faces of A, once for each point where the ray passes from the interior to the exterior of A or vice-versa. Thus the predicate for containment of a point $\mathbf{p}_B$ in A is the xor of edge-face predicates for the ray e_B and all faces of A. i.e.

$$O_{p_B,A} \;=\; \bigotimes_{\substack{f_i \in \\ \mathrm{faces}(A)}} O_{e_B,f_i} \tag{2.8}$$

where O_{e_B,f_i} is given by equations (2.4), (2.5) and (2.6). Since e_B has a head but no tail, the predicate $A_{T,f}$ in (2.5) and (2.6) is not defined. However, since conceptually the tail is infinitely far away in direction $-d_B$, $A_{T,f}$ should be set to true iff the inner product of the direction d_B with the outward normal to f is *negative.*

Even if the ray direction d_B is generic, there are some degenerate placements of A and B which give ambiguous values for the number of faces that intersect the ray (such as where the ray intersects the edge between two faces, in some cases, this should count as two intersections, in other cases, one). However, for the collision detection algorithm described later, degeneracies merely show up as point "glitches" along a path, where the overlap predicate suddenly goes from overlap to non-overlap and back again. They are removed by a robust interval list merging algorithm.

Another way to eliminate degenerate ray positions is to use several ray directions from the same point $\mathbf{p}_B$. If the ray directions are chosen generically, at most three of the rays will intersect edges of A, no matter how A and B are placed. So choosing seven or more generic rays guarantees that there are always at least four of them that do not intersect edges of A, and these four give the correct value for inside/outside. After defining inside/outside predicates for each of these rays via equation (2.8), we can then combine their outputs with a binary "consensus" function of seven inputs, i.e. a function which is true if and only if four or more of its inputs are true.

This predicate would always give the correct indication of inside/outside for the point $\mathbf{p}_B$ in A.

Finally, we observe that the number of primitive predicates appearing in the overlap predicate (2.1) is proportional to the product of the number of features of A and the number of features of B. To see this we notice that the point-polyhedron predicate (2.8) enumerates faces of A and then edges of each face. But every edge of A is enumerated only twice in this way. The edge-face comparison requires enumeration over edges of A, faces of B and then edges of each face of B. But again the edges of B only appear twice (for each edge of A) in this enumeration.

2.1.2 Conjunctive Form

The conjunctive form of overlap predicate for two polyhedra is really based on a condition for non-overlap. Like the disjunctive form, it is built from the type A and type C primitive predicates, which involve vertex-face or edge-edge interactions respectively.

Definition We define the *wedge* of an edge e of A as the intersection of the two interior half-spaces of the faces which cobound e. The wedge of e is denoted $\hat{e}$, and it contains A.

Lemma 2.1.2 *Two convex polyhedra A and B are non-overlapping iff either* (i) *all edges of A are outside some wedge of an edge of B, or* (ii) *all edges of B are outside some wedge of an edge of A.*

Proof. This condition is clearly sufficient for non-overlap, by convexity of A and B. Conversely, if A and B are disjoint, then there is a (not necessarily unique) non-zero shortest vector between them. Let p_A and p_B be the end-points of this vector in A and B respectively. If one of these points lies in the interior of a face f, then the test succeeds for any wedge of an edge in the boundary of f. If one of the points lies in the interior of an edge e, then the test succeeds for $\hat{e}$

The only case remaining is where p_A and p_B are both vertices. Let f be a face adjacent to p_A, and such that p_B lies outside the interior half-space $\bar{f}^-$, (there must be at least one such f, or p_B would be contained in A).

Then if B is also outside this half-space, the test succeeds for the wedge of any edge that cobounds f.

Otherwise, let W_{p_B} be the intersection of the wedges of all edges that cobound p_B. Let S be the plane passing through p_B and normal to the vector $(p_B - p_A)$. S defines two closed half-spaces, one (S_A) containing A and the other (S_B) containing both B and W_{p_B}, i.e. S separates A and B. Then the intersection $(W_{p_B} \cap \bar{f})$ lies in $(S_B \cap \bar{f})$. Let p_0 be the closest point in $(W_{p_B} \cap \bar{f})$ to $(S \cap \bar{f})$. Then p_0 is in the boundary of some wedge $\hat{e}$ of an edge e that cobounds p_B. Now $(\hat{e} \cap \bar{f}) \subset (S_B \cap \bar{f})$ and so by projection from p_B, $(\hat{e} \cap \bar{f}^-) \subset (S_B \cap \bar{f}^-)$. But $A \subset (S_A \cap \bar{f}^-)$, so $(\hat{e} \cap A) = \emptyset$, and the test succeeds for $\hat{e}$. □

Thus we can define the following predicate for non-overlap $F_{A,B}$ of A and B from the above test:

$$F_{A,B} = \Big(\bigvee_{\substack{e_j \in \\ \text{edges}(B)}} \bigwedge_{\substack{e_i \in \\ \text{edges}(A)}} F_{\hat{e}_j, e_i} \Big) \vee \Big(\bigvee_{\substack{e_i \in \\ \text{edges}(A)}} \bigwedge_{\substack{e_j \in \\ \text{edges}(B)}} F_{\hat{e}_i, e_j} \Big) \tag{2.9}$$

where $F_{\hat{e}_i, e_j} = ((\hat{e}_i \cap e_j) = \phi)$. If the object consists of several convex pieces A_i, as do the obstacles B_j, then the non-overlap predicate is the conjunction of pairwise predicates

$$F = \bigwedge_i \bigwedge_j F_{A_i, B_j} \tag{2.10}$$

where each F_{A_i,B_j} is given by equation (2.9). Now we can define $F_{\hat{e}_A, e_B}$ in terms of the above predicates. Let L and R be the left and right faces respectively, which cobound e_A, (left and right are determined here by viewing e_A from outside $\hat{e}_A$ with $\mathbf{d}_A$ upward).

Lemma 2.1.3 *The following predicate indicates non-overlap of the wedge $\hat{e}_A$ and the edge e_B.*

$$\begin{aligned} F_{\hat{e}_A, e_B} \;=\; & ((A_{H,L} \wedge (A_{T,L} \vee (A_{T,R} \wedge C^+_{e_A, e_B}))) \vee \\ & (A_{H,R} \wedge (A_{T,R} \vee (A_{T,L} \wedge C^-_{e_A, e_B})))) \end{aligned} \tag{2.11}$$

Proof The proof is by case analysis of all possible primitive predicate values. Firstly, the predicate $A_{H,L}$ requires the head of e_B to be above the plane of the left face of e_A. For the following subcases, we assume that $A_{H,L}$ is true, so that H is above the plane of L. The subcases are itemized below:

- If $A_{T,L}$ is true, then the tail T of e_B is also above the plane of L, i.e. the entire edge e_B is outside of $\hat{e}_A$ and the predicate correctly returns true.

- If $A_{T,L}$ is false, and so is $A_{T,R}$, then the vertex T lies inside $\hat{e}_A$, and the predicate $F_{\hat{e}_A,e_B}$ correctly returns false.

- The only case remaining is where $A_{T,L}$ is false and $A_{T,R}$ is true, so that both vertices are outside of $\hat{e}_A$, and H is above the plane of L and T is below that plane. This case is illustrated in figure 2.1. In this case $d_A \times d_B$ is a vector which points out of $\hat{e}_A$, and is normal to d_A and d_B. Then e_A is outside of $\hat{e}_A$ if and only if the inner product of $(H - p_A)$ with this vector is positive, a condition which is indicated by $C^+_{e_A,e_B}$.

We have shown that the predicate returns correct values if $A_{H,L}$ is true. A similar analysis for the second line of expression (2.11) shows it returns correct values if $A_{H,R}$ is true. The only case remaining is where both $A_{H,L}$ and $A_{H,R}$ are false, but here the point H is inside the object and the predicate $F_{\hat{e}_A,e_B}$ correctly returns a false value. □

With some rearrangement, the predicate $F_{\hat{e}_A,e_B}$ can be written in the equivalent form:

$$\begin{aligned} F_{\hat{e}_A,e_B} = & (A_{H,L} \vee A_{H,R}) \wedge (A_{T,L} \vee A_{T,R}) \wedge \\ & (A_{H,L} \vee A_{T,R} \vee C^+_{e_A,e_B}) \wedge (A_{H,R} \vee A_{T,L} \vee C^-_{e_A,e_B}) \end{aligned} \tag{2.12}$$

There is a similar form for the predicate $F_{\hat{e}_B,e_A}$.

Finally, the static *overlap* predicate we are seeking is the negation of the non-overlap predicate (2.9), and has the following form, with an AND-node at its root:

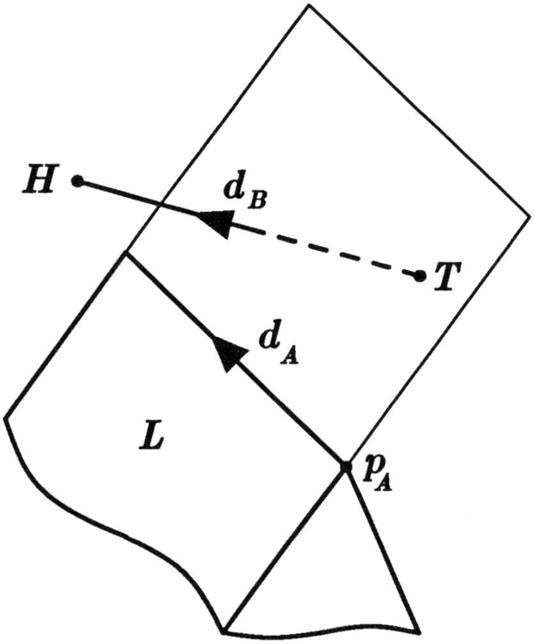

Figure 2.1: Position of edge e_B outside the wedge $\hat{e}_A$, where inside/outside is determined by $C^+_{e_A,e_B}$

$$O_{A,B} = \Big(\bigwedge_{\substack{e_j \in \\ \text{edges}(B)}} \bigvee_{\substack{e_i \in \\ \text{edges}(A)}} \neg F_{\hat{e}_j,e_i} \Big) \bigwedge \Big(\bigwedge_{\substack{e_i \in \\ \text{edges}(A)}} \bigvee_{\substack{e_j \in \\ \text{edges}(B)}} \neg F_{\hat{e}_i,e_j} \Big) \qquad (2.13)$$

In defining the overlap predicate we have not made use of explicit "applicability constraints" for the constraint surfaces [Loz], and in detail in [Don]. The applicability constraints delimit ranges of rotations for which a particular type of contact can occur. The configuration space obstacle between a convex object and obstacles can be expressed as a conjunction of applicable constraints (each of which is the disjunction of a constraint and its applicability constraints), so it is closer to the conjunctive form we derived above.

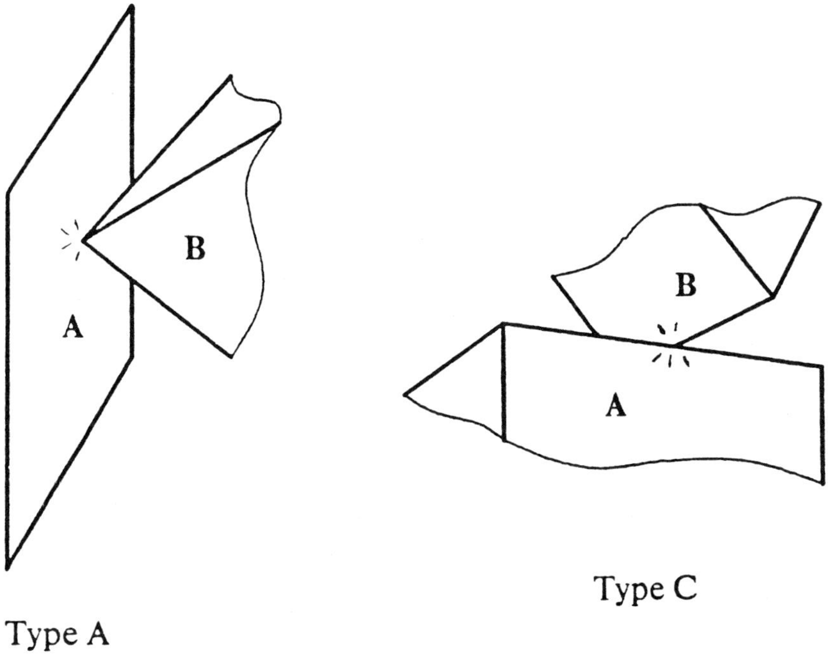

Figure 2.2: The two types of contact between polyhedra

2.2 Contact Conditions

We have seen that both conjunctive and disjunctive overlap predicates are defined in terms of two primitive predicates, which we have denoted $A_{p,f}$ for a point **p** and face f, and C_{e,e_i} for edges e and e_i. These predicates correspond to the two types of contact that can occur between polyhedra when they touch. Lozano-Pérez [Loz] first defined constraints in configuration space from these contact conditions. He used the term type A for the contact that occurs when a vertex of one polyhedron touches a face of the other as shown in figure 2.2. Type C contact occurs when an edge of one polyhedron touches an edge of the other. In fact, Lozano-Pérez distinguishes two types of vertex-face contact, depending on whether the vertex belongs to a polyhedron which is designated as part of the robot or whether it belongs to a fixed obstacle. However, in the more general case, where both polyhedra may be moving (e.g. two different links on a robot arm) this distinction cannot be made. We will only make the distinction in section 2.3, where we compute the constraints for a single moving polyhedron. Otherwise, it is enough to ensure that all tests are performed symmetrically on both polyhedra.

In general, we can describe the position and orientation of all the links in a robot (or multi-robot system) by a configuration x which lies in a configuration space H. We will assume that H is an algebraic subset of some $\Re^n$. This assumption is valid for virtually all robot kinematics. Thus the coordinates of vertices become functions of x, as do homogeneous face normals and edge directions. In turn, the predicates $A_{p,f}$ and C_{e,e_i} become functions of x, and to derive their new form, we simply substitute the transformed features $\mathbf{p}(x)$, $\mathbf{n}(x)$, $c(x)$ and $\mathbf{d}(x)$. This gives us the overlap predicates which define the configuration space obstacle. Later in this chapter, we give explicit derivation of these predicates for the case of an unconstrained polyhedron, which has three positional and three rotational degrees of freedom, and for robots with rotary joints. First we describe the free polyhedron case.

2.3 Constraints for Free Polyhedra

Since the representation of the configuration has a dramatic effect on the running time of the algorithms that follow, we must be careful to choose the most efficient parameterization possible. Quaternions seem to be the best choice for representing the orientation of a free polyhedron, and so we next review some of their basic properties.

2.3.1 Review of Quaternions

We will be using quaternions to describe rotations of vectors in Euclidean 3-space $\Re^3$. By convention, symbols representing vectors will be lower case bold ($\mathbf{u}$, $\mathbf{v}$), quaternions will upper case bold ($\mathbf{P}$, $\mathbf{Q}$) and scalars set in lower case. A quaternion $\mathbf{Q} \in \Re^4$ can be split into a scalar part and a 3-vector. We denote the scalar part of the quaternion $\mathbf{Q}$ by q_0 and the vector part by $\mathbf{q}$, and to simplify notation, the scalar and vector part will be combined using an addition symbol, so that $\mathbf{Q} = q_0 + \mathbf{q}$. It will also be convenient to treat vectors as quaternions having zero scalar part, and scalars as quaternions with zero vector part.

Quaternions form a 4-dimensional algebra, i.e. vector space with an associative multiplication. The product $\mathbf{R} = \mathbf{PQ}$ is given by

$$\begin{aligned} r_0 &= p_0 q_0 - \mathbf{p} \cdot \mathbf{q} \\ \mathbf{r} &= p_0 \mathbf{q} + q_0 \mathbf{p} + \mathbf{p} \times \mathbf{q} \end{aligned} \tag{2.14}$$

The product is equivalent to scalar multiplication if one of $\mathbf{P}$ or $\mathbf{Q}$ has zero vector part. i.e. $\alpha\mathbf{Q}$ gives the same result whether we interpret the product as scalar multiplication or as a quaternion product where α represents the quaternion $\alpha + \mathbf{0}$. So we find that $\alpha\mathbf{Q} = \mathbf{Q}\alpha$ even though quaternion product is not commutative in general.
Notice that quaternion product is linear in $\mathbf{P}$ and $\mathbf{Q}$:

$$\mathbf{P}(\mathbf{Q} + \lambda\mathbf{S}) = \mathbf{PQ} + \lambda\mathbf{PS}$$

The conjugate $\mathbf{Q}^*$ of a quaternion $\mathbf{Q}$ is formed by negating the vector part of $\mathbf{Q}$, so that

$$\mathbf{Q}^* = q_0 - \mathbf{q}$$

The product $\mathbf{QQ}^* = q_0^2 + \mathbf{q} \cdot \mathbf{q}$ produces a scalar (quaternion with zero vector part) which is the squared magnitude of $\mathbf{Q}$, i.e.

$$|\mathbf{Q}| = \sqrt{\mathbf{QQ}^*}$$

is the length of $\mathbf{Q}$ in $\Re^4$ under the Normal Euclidean metric. If $\mathbf{Q}$ is a unit quaternion, it satisfies $\mathbf{QQ}^* = 1$, so that $\mathbf{Q}^*$ is the multiplicative inverse of $\mathbf{Q}$.

There is more than one way to represent a rotation with a quaternion, but the most common, and the one we will be using, has a simple expression in terms of the quaternion product. Suppose we rotate the vector $\mathbf{v}$ about an axis $\mathbf{n}$ through an angle of θ, giving $\mathbf{v}'$. The rotated vector $\mathbf{v}'$ is given by

$$\mathbf{v}' = \mathbf{Q}\mathbf{v}\mathbf{Q}^* \tag{2.15}$$

where

$$\mathbf{Q} = \cos\frac{\theta}{2} + \mathbf{n}\sin\frac{\theta}{2} \tag{2.16}$$

With this representation, every rotation is specified by a unit quaternion $\mathbf{Q}$, and these form a group under quaternion product. Composition of rotations requires a single quaternion product, i.e. a rotation of $\mathbf{Q}$ followed by a rotation of $\mathbf{P}$ is the quaternion $\mathbf{PQ}$. Inverse rotations are found by conjugation, so the inverse of $\mathbf{Q}$ is $\mathbf{Q}^*$. The representation defines a homomorphism from the space of unit quaternions S^3 to the group $SO(3)$ of rotations in $\Re^3$. It is not an isomorphism since $\mathbf{Q}$ and $-\mathbf{Q}$ define the same rotation.

Quaternion product provides a very convenient shorthand for various operations on vectors, and we will use it extensively in defining motion constraints. It will often be necessary to extract the scalar part of a quaternion $\mathbf{Q}$ which we do by surrounding it with square brackets

$$[\mathbf{Q}] = q_0$$

Then for vectors $\mathbf{u}$, $\mathbf{v}$, $\mathbf{w}$, we have compact forms for dot product and scalar triple product:

$$\begin{aligned} [\mathbf{uv}] = [\mathbf{vu}] &= -\mathbf{u}\cdot\mathbf{v} \\ [\mathbf{uvw}] = [\mathbf{vwu}] &= -\mathbf{u}\cdot\mathbf{v}\times\mathbf{w} \end{aligned} \tag{2.17}$$

We will also make use of the fact that $[\mathbf{Qx}] = [(q_0 + \mathbf{q})\mathbf{x}] = [\mathbf{qx}]$, i.e. that the scalar part of a vector-quaternion product depends only on the vector part of the quaternion.

Since we wish to use quaternions to represent rotations, we are really interested in unit quaternions, in fact in the projective space P^3 obtained by identifying quaternions which are related by scaling. This space is three-dimensional, but cannot be embedded in three dimensional affine space $\Re^3$. However, it can be represented as a union of finitely many affine pieces. The following "four-cubes" representation seems to be the most efficient choice for representing the orientation of a free object.

We can represent rotations using four 3-dimensional cubes C_i, $i = 1, \ldots, 4$ in $\Re^4$, the space of quaternion coordinates. Every unit quaternion is the projection of a quaternion in some C_i. To find the equivalent quaternion in a cube, take any unit quaternion, and divide each coordinate by the coordinate with largest magnitude, say the ith coordinate. Then the resulting quaternion has $q_i = 1$ and $|q_j| \leq 1$ for $j \neq i$, and lies in the ith cube. A unit quaternion may map into more than one cube only if two or more of its coordinates have the maximum magnitude. In this case it maps onto either the faces of two distinct cubes (if two coordinates have maximum magnitude), the edges of three distinct cubes (three coordinates of maximum magnitude), or the vertices of all four cubes (four coordinates of the same magnitude). This allows a unique identification of the faces of the cubes and yields a 3-dimensional topological space which is homeomorphic to P^3 and to $SO(3)$, the group of three dimensional rotations.

We can use this representation directly if the constraint equations are homogeneous and of even degree in the quaternion coordinates. This is because the sign of a homogeneous polynomial of even degree is unaffected by scaling. If this is the case, then the sign of any constraint equation is the same whether we compute its value using a unit quaternion or the representative of that unit quaternion in one of the cubes. For collision detection, we will be able to make use of a straight line path in $\Re^4$, and the results will be the same as would be obtained if the path were projected onto the sphere of unit quaternions.

For more applications of quaternions in robotics the reader should see [PW], [Sa] and [Yan] , and the most complete reference on the theory of quaternions is Hamilton's original work [Ham].

2.3.2 Transformation of Object Features

Here we consider the result of rotating and translating each type of feature on the object polyhedron, in preparation for the derivation of motion constraints. The three types of feature that we need to consider are:

1. Vertices, specified by the vector from the origin to the vertex e.g. $\mathbf{p}_A$.

2. Edges, specified by a unit vector in the edge direction $\mathbf{d}_A$, and by a vertex lying on the edge $\mathbf{p}_A$.

3. Faces, specified by a unit outward normal vector $\mathbf{n}_A$, and the normal distance from the origin to the face $c_A = \mathbf{n}_A \cdot \mathbf{p}_A$ for any vertex $\mathbf{p}_A$ lying on the face.

Alternatively, we can represent a face using the quaternion $\mathbf{N}_A = c_A + \mathbf{n}_A$, usually called the homogeneous normal. This simplifies the equation of the plane through the face. A point $\mathbf{y}$ lies in the plane through $\mathbf{N}_A$ iff

$$\mathbf{y} \cdot \mathbf{n}_A - c_A = -[\mathbf{N}_A(1 + \mathbf{y})] = 0 \tag{2.18}$$

This form is also very convenient for spatial transformations, as we shall see in a moment.

In our configuration space, rotations must be performed before translation, so all the features of A must first be rotated by $\mathbf{Q}$. If $\mathbf{G} \in \{\mathbf{p}_A, \mathbf{d}_A, \mathbf{N}_A\}$ is any feature, and let $\mathrm{rot}_Q(\mathbf{G})$ be the rotated feature, then

$$\mathrm{rot}_Q(\mathbf{G}) = \mathbf{Q}\mathbf{G}\mathbf{Q}^* \tag{2.19}$$

This is clear for vertices and edge directions which are vectors. For faces, the outward normal should be rotated by $\mathbf{Q}$ while the distance c_A should be unaffected. This follows from

$$\mathrm{rot}_Q(\mathbf{N}_A) = \mathbf{Q}(c_A + \mathbf{n}_A)\mathbf{Q}^* = c_A + \mathbf{Q}\mathbf{n}_A\mathbf{Q}^*$$

which is the correctly rotated homogeneous normal.

Let $\mathrm{Trans}_x(\mathbf{G})$ be translation of the feature $\mathbf{G}$ by $\mathbf{x}$. Unlike rotation, this transformation has a different effect on each type of feature.

- Edge directions remain unchanged by translation, so $\mathrm{Trans}(\mathbf{d}_A) = \mathbf{d}_A$.

- Vertices are displaced by $\mathbf{x}$, which requires only vector addition, $\mathrm{Trans}_x(\mathbf{p}_A) = \mathbf{p}_A + \mathbf{x}$.

- $\mathrm{Trans}_x(\mathbf{N}_A) = \mathbf{N}_A - [\mathbf{N}_A\mathbf{x}]$, the proof follows:

Face outward normals are unchanged by translation, so $\mathrm{Trans}_x(\mathbf{n}_A) = \mathbf{n}_A$. However c_A is defined as the dot product of $\mathbf{n}_A$ with some vertex $\mathbf{p}_A$ that lies on the face, which will move during translation, therefore $\mathrm{Trans}_x(c_A) = \mathbf{n}_A \cdot \mathrm{Trans}_x(\mathbf{p}_A)$. Substituting for the displaced vertex, we obtain

$$\mathrm{Trans}(c_A) = \mathbf{n}_A \cdot \mathbf{p}_A + \mathbf{n}_A \cdot \mathbf{x} = c_A + \mathbf{n}_A \cdot \mathbf{x}$$

Using the identities in (2.17), we can write $\text{Trans}(c_A) = c_A - [\mathbf{n}_A\mathbf{x}]$. The overall transformation of $\mathbf{N}_A$ is

$$\text{Trans}(\mathbf{N}_A) = \mathbf{N}_A - [\mathbf{n}_A\mathbf{x}] = \mathbf{N}_A - [\mathbf{N}_A\mathbf{x}]$$

Summarizing these transformations:

$$\begin{array}{ll} \text{Trans}_x(\mathbf{p}_A) = \mathbf{p}_A + \mathbf{x}, & \text{for vertices} \\ \text{Trans}_x(\mathbf{d}_A) = \mathbf{d}_A, & \text{for edge directions} \\ \text{Trans}_x(\mathbf{N}_A) = \mathbf{N}_A - [\mathbf{N}_A\mathbf{x}], & \text{for faces} \end{array} \tag{2.20}$$

The transformed feature $\mathbf{G}'$, when A is at configuration $(\mathbf{x}, \mathbf{Q})$ is $\mathbf{G}' = \text{Trans}_x(\text{rot}_Q(\mathbf{G}))$. Substituting from equations (2.19) and (2.20), we have

$$\begin{aligned} \mathbf{p}'_A &= \mathbf{Q}\mathbf{p}_A\mathbf{Q}^* + \mathbf{x} \\ \mathbf{d}'_A &= \mathbf{Q}\mathbf{d}_A\mathbf{Q}^* \\ \mathbf{N}'_A &= \mathbf{Q}\mathbf{N}_A\mathbf{Q}^* - [\mathbf{Q}\mathbf{N}_A\mathbf{Q}^*\mathbf{x}] \end{aligned} \tag{2.21}$$

for the new features after rotation by $\mathbf{Q}$ and translation by $\mathbf{x}$.

2.3.3 Primitive Predicates

To obtain the type-A constraints for transformed features, we substitute for faces of A from (2.21) into (2.3), giving:

$$A_{p_B, f_A} = ([\mathbf{Q}\mathbf{N}_A\mathbf{Q}^*(1 + \mathbf{p}_B - \mathbf{x})] > 0) \tag{2.22}$$

Which is already homogeneous and of even degree in the quaternion coordinates. Similar expressions arise for interaction between a face of B and a vertex of A. (This is sometimes called type-B contact).

$$A_{p_A, f_B} = ([\mathbf{N}_B(1 + \mathbf{Q}\mathbf{p}_A\mathbf{Q}^* + \mathbf{x})] > 0) \tag{2.23}$$

Notice that this polynomial is of even degree, but not homogeneous in the quaternion coordinates. However, there is a simple modification that we can make to ensure this, which does not change its value for unit quaternions.

$$A_{p_A, f_B} = ([\mathbf{N}_B(\mathbf{Q}\mathbf{p}_A\mathbf{Q}^* + \mathbf{Q}\mathbf{Q}^*(1 + \mathbf{x}))] > 0) \tag{2.24}$$

This is the form we will make use of in future. The type-C contraints (2.7) have the form

$$C^+_{e_A,e_B} = ([(\mathbf{Q}\mathbf{p}_A\mathbf{Q}^* + \mathbf{x} - \mathbf{p}_B)\mathbf{Q}\mathbf{e}_A\mathbf{Q}^*\mathbf{e}_B] > 0) \tag{2.25}$$

where $\mathbf{p}_A$ and $\mathbf{p}_B$ are points on e_A and e_B respectively, and $\mathbf{d}_A$ and $\mathbf{d}_B$ are their respective edge directions. After some rearrangement, using linearity of quaternion product, and removing $\mathbf{Q}\mathbf{Q}^*$ products, we obtain an expression of type-C constraint with lower algebraic degree in the quaternion coordinates.

$$C^+_{e_A,e_B} = ([\mathbf{Q}\mathbf{p}_A\mathbf{e}_A\mathbf{Q}^*\mathbf{e}_B] + [(\mathbf{x} - \mathbf{p}_B)\mathbf{Q}\mathbf{e}_A\mathbf{Q}^*\mathbf{e}_B] > 0) \tag{2.26}$$

All the constraints (2.22), (2.24), (2.26) are algebraic, that is, they are polynomial in the coordinates of $\mathbf{x}$ and $\mathbf{Q}$. In fact they are all linear in $\mathbf{x}$ and homogeneous quadrics in $\mathbf{Q}$.

2.4 An Algorithm for Collision Detection

If instead of arbitrary values for $\mathbf{x}$ and $\mathbf{Q}$ we have functions $\mathbf{x}(s)$ and $\mathbf{Q}(s)$ of some parameter s, the predicates (2.9) and (2.1) become binary-valued functions of s. The true-false transitions of these functions and of any of the predicates $A_{p,f}(s), C^+_{e_A,e_B}(s)$ correspond to zero-crossings of one of the expressions (2.22), (2.24), (2.26) as functions of the parameter s. Finding the zero-crossings of one of these functions gives us the free path segments for the corresponding primitive predicate, and we can find the free segments of the overlap predicate by computing unions and intersections of primitive predicate path segments.

If the functions $\mathbf{x}(s)$ and $\mathbf{Q}(s)$ are polynomial in the variable s, finding the constraint zero-crossings is particularly simple. When we substitute for $\mathbf{x}$ and $\mathbf{Q}$ in equations (2.22), (2.24) and (2.26), we obtain *univariate* polynomials in s, and we can apply standard techniques to find their zeros [CL]. By considering a particular path through configuration space we have reduced a problem involving polynomials in six variables to one involving polynomials in one variable. The simplest possible polynomial path is a straight line in configuration space, which corresponds to linear motion and approximately uniform rotation in real space. An implementation of a collision detector for linear paths is described next.

Let us assume that before the motion, the object is at the "origin" of configuration space, i.e. that $\mathbf{x} = (0,0,0)$, $\mathbf{Q} = (1,0,0,0)$. There is no loss of generality since we can always define vertex positions, plane equations etc. for the object at its start position. The motion of the object is a straight line in configuration space from the origin to the final configuration $(\mathbf{x}_f, \mathbf{Q}_f)$, where $\mathbf{Q}_f = 1 + \mathbf{q}_f$. This path can be made a function of a single parameter s, and intermediate configurations are of the form $(s\mathbf{x}_f, 1 + s\mathbf{q}_f)$. When we substitute this configuration into (2.22), (2.24) and (2.26) we find that all three constraints reduce to univariate cubics in the parameter s. If we take s to be a time parameter, the motion of the object is uniform in the following sense. From (2.16), the vector part of the quaternion having $q_0 = 1$ can be written $\mathbf{q} = \mathbf{n} \tan \frac{\theta}{2}$. The orientation of the object along the path can also be written in this form

$$s\mathbf{q}_f = \hat{\mathbf{q}}_f \tan \frac{\theta}{2} \tag{2.27}$$

where $\hat{\mathbf{q}}_f = \frac{\mathbf{q}_f}{|\mathbf{q}_f|}$. So the object is rotating about a fixed axis $\hat{\mathbf{q}}_f$, but the angular velocity is not constant because θ does not vary linearly with s. They are related instead by

$$\theta = 2 \operatorname{atan}(s|\mathbf{q}_f|) \tag{2.28}$$

The object always begins with angular velocity of $2|\mathbf{q}_f|$, which is the initial rate of change of θ with s, and decelerates smoothly. For small $\mathbf{q}_f$ the change in velocity is small, and the final angular velocity is

$$\frac{2|\mathbf{q}_f|}{1 + |\mathbf{q}_f|^2} \tag{2.29}$$

So we find that a straight line path through configuration space gives us a "natural" motion through real space, that is, constant linear velocity and nearly constant angular velocity.

Substituting the configuration $(s\mathbf{x}_f, s\mathbf{q}_f)$ into the type A equation (2.22), we get

$$A_{p_B,f_A} = ([(1 + s\mathbf{q}_f)\mathbf{N}_A(1 - s\mathbf{q}_f)(1 + \mathbf{p}_B - s\mathbf{x}_f)] > 0) \tag{2.30}$$

and rearranging in powers of s we obtain:

$$\begin{aligned} A_{p_B,f_A}(s) = \; & s^0[\mathbf{N}_A(1 + \mathbf{p}_B)]+ \\ & s^1[(\mathbf{q}_f\mathbf{N}_A - \mathbf{N}_A\mathbf{q}_f)(1 + \mathbf{p}_B) - \mathbf{N}_A\mathbf{x}_f]+ \\ & s^2[-\mathbf{q}_f\mathbf{N}_A\mathbf{q}_f(1 + \mathbf{p}_B) - (\mathbf{q}_f\mathbf{N}_A - \mathbf{N}_A\mathbf{q}_f)\mathbf{x}_f]+ \\ & s^3[\mathbf{q}_f\mathbf{N}_A\mathbf{q}_f\mathbf{x}_f] \end{aligned} \tag{2.31}$$

Since $A_{p_B,f_A}(s)$ is cubic in s, its zeros can be easily found, and at each simple zero, $A_{p_B,f_A}(s)$ makes a true-false transition. There can be at most 3 transitions in the interval $[0,1]$ so the set, $\{s \in [0,1] | A_{p_B,f_A}(s)\}$, consists of at most two intervals.

More generally, we can represent any predicate as the list of intervals in $[0,1]$ in which the predicate is true. Disjunctions and conjunctions of predicates can be found by a kind of merge operation on lists. e.g. suppose the list l_1 represents the predicate $P_1(s)$ and l_2 represents $P_2(s)$. The list l_3 representing $P_1(s) \wedge P_2(s)$ can be computed in linear time in the total input size $|l_1| + |l_2|$, assuming the interval lists are sorted:

algorithm intersect;
input $l_1 = \langle [a_0, b_0] \ldots [a_n, b_n] \rangle$ **and** $l_2 = \langle [c_0, d_0] \ldots [c_m, d_m] \rangle$;
$i \leftarrow 0$; $j \leftarrow 0$;
repeat
if $b_i < c_j$ **then** $i \leftarrow i + 1$; — no overlap; skip an interval in l_1
  else if $d_j < a_i$ **then** $j \leftarrow j + 1$; — no overlap; skip an interval in l_2
    else begin; — overlap; intersect current intervals
      $l_3 \leftarrow l_3 \circ$ [**max**(a_i, c_j), **min**(b_i, d_j)];
      if $b_i < d_j$ **then** $i \leftarrow i + 1$;
        else $j \leftarrow j + 1$;
      end;
  until $i > n$ **or** $j > m$;
output l_3;

where "o" denotes concatenation. The algorithm for interval union is similar, and also runs in linear time. We note here that both intersection and union operations produce a list l_3 with at most $|l_1| + |l_2|$ intervals. Using equations (2.9) or (2.1), we perform intersection and union of interval lists until we arrive at an overlap predicate. The interval list for this predicate describes the points at which overlap actually occurs.

The fastest way to merge all the constraints, using a divide and conquer strategy, is to first merge pairs of constraints, then taking the interval lists output at this "phase" and merging pairs of them etc. The time required for each phase is linear in the *total* number of intervals of all predicates being merged in that phase. If we start with N_c constraints, the total number of intervals before the first merge is $2N_c$. This number does not increase with successive merges, because merging N lists of length m produces $\frac{N}{2}$ lists of length at most $2m$. Merging is complete after $\log(N_c)$ phases, giving an overall complexity of $O(N_c \log N_c)$, in terms of the number of constraints. This time dominates the time to compute the zeros of the constraint polynomials, which is $O(N_c)$.

In terms of the complexity of the environment, N_c is given by $N_c = E_A E_B + F_A V_B + F_B V_A$ where F_A, V_A and E_A are respectively the number of faces, vertices and edges of the object A, and F_B, V_B and E_B are the corresponding numbers for the obstacles. A reasonable measure of the input size is $n = V_A + E_A + F_A + V_B + E_B + F_B$, and the time complexity is then $O(n^2 \log n)$. While in a strict sense, the algorithm takes pseudo-quadratic time in terms of its input, this worst case arises only if both the object and obstacles increase in complexity with n. If the complexity of the object is fixed, the algorithm runs in pseudo-linear time as a function of the complexity of the obstacles.

An example of the operation of the collision detector is given in figure 2.3. In this figure the L-shaped object moves and rotates uniformly between the configurations labelled C_i and C_f. The collision detector returns a series of collision points which correspond to the object and obstacle just starting to touch. These are illustrated from a different angle in the next frame. The algorithm was implemented on a Symbolics Lisp Machine and takes about 15 seconds for the problem shown.

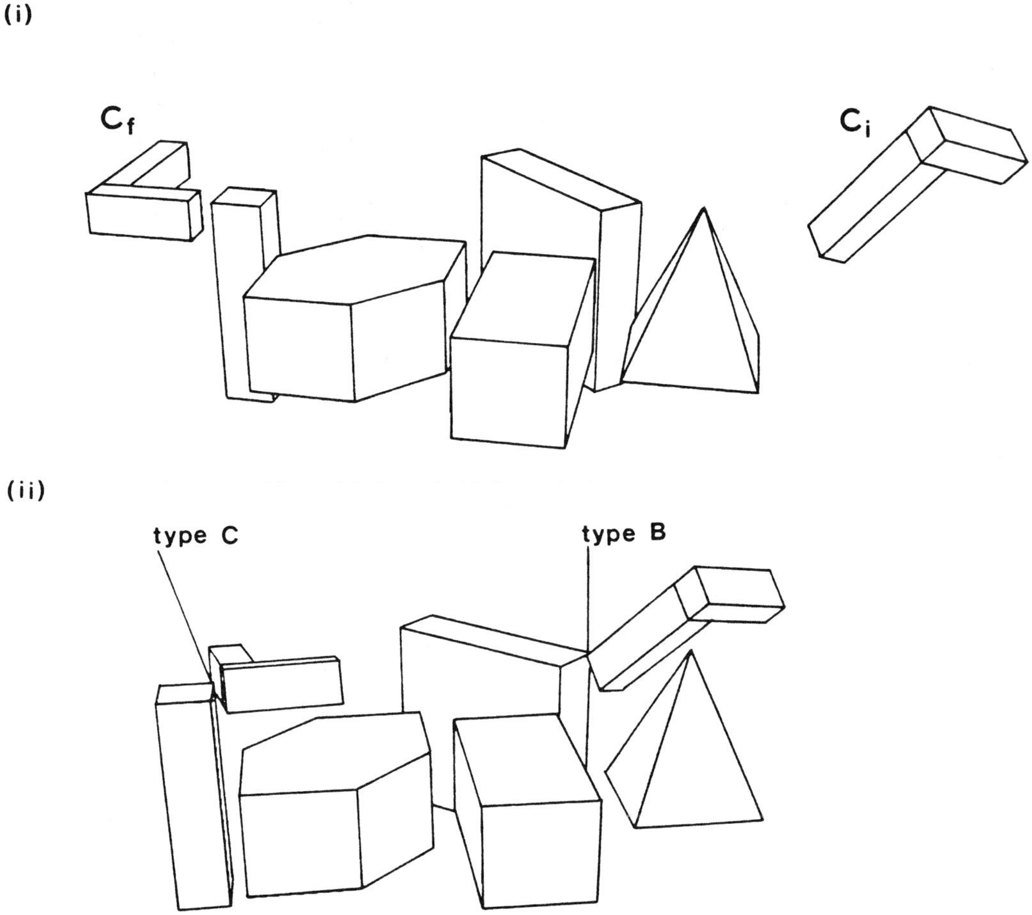

Figure 2.3: Collision detection example, (i) L-shaped object moves from configuration C_i to C_f, (ii) Collision points and contact types along the path.

2.5 Collision Detection for Several Moving Objects

So far we have assumed that we are given a path for a single object moving among fixed obstacles. We can readily extend the interference test to deal with several moving objects by testing pairwise for interference between objects. The configuration space for 2 objects in motion is 12-dimensional, but note that the expressions for overlap in equations (2.9), (2.1) are valid independent of the configuration space. When the trajectory of both objects is known, the configuration parameters are still all functions of a single argument, so once again collision detection reduces to solving univariate polynomials. The only significant difference is in the constraint equations themselves, and we now give the general form for A, B and C constraints between moving objects.

We denote one of the moving objects A and the other B, and let the configuration of A be $(\mathbf{x}_A, \mathbf{Q}_A)$, and that of B be $(\mathbf{x}_B, \mathbf{Q}_B)$ (we will assume normalized quaternions for the moment). Then let p'_B and f'_A denote the transformed features. Recall that the equation for type-A interaction with B fixed is

$$A_{p_B, f'_A} = ([\mathbf{Q}\mathbf{N}_A\mathbf{Q}^*(1 + \mathbf{p}_B - \mathbf{x})] > 0) \tag{2.32}$$

But if B has configuration $(\mathbf{x}_B, \mathbf{Q}_B)$, then $\mathbf{p}'_B$ will be $\mathrm{rot}(\mathbf{p}_B, \mathbf{Q}_B) + \mathbf{x}_B$, and the generalized type-A constraint is

$$A_{p'_B, f'_A} = ([\mathbf{Q}_A\mathbf{N}_A\mathbf{Q}_A^*(1 + \mathbf{Q}_B\mathbf{p}_B\mathbf{Q}_B^* + \mathbf{x}_B - \mathbf{x}_A)] > 0) \tag{2.33}$$

Again we multiply certain terms by $\mathbf{Q}\mathbf{Q}^*$ to ensure that the constraint polynomial is homogeneous, and we do this independently for $\mathbf{Q}_A$ and $\mathbf{Q}_B$. We first make equation (2.33) independent of quaternion magnitude. The scale independent form is

$$A_{p'_B, f'_A} = ([\mathbf{Q}_A\mathbf{N}_A\mathbf{Q}_A^*\,(\mathbf{Q}_B\mathbf{p}_B\mathbf{Q}_B^* + \mathbf{Q}_B\mathbf{Q}_B^*(1 + \mathbf{x}_B - \mathbf{x}_A))] > 0) \tag{2.34}$$

If both A and B are moving, type A and B constraints are of the same form, since both arise when a vertex of one object touches a face of the other.

For type C constraints we take equation (2.7) for edge-edge interaction for a stationary object B

$$C^{+}_{e_A,e_B} = ([\mathbf{Q}\mathbf{p}_A\mathbf{e}_A\mathbf{Q}^*\mathbf{e}_B] + [(\mathbf{x} - \mathbf{p}_B)\mathbf{Q}\mathbf{e}_A\mathbf{Q}^*\mathbf{e}_B] > 0) \tag{2.35}$$

and rotate the edge $\mathbf{e}_B$ by $\mathbf{Q}_B$, and rotate and translate the vertex $\mathbf{p}_B$, giving

$$\begin{aligned} C^{+}_{e_A,e_B} = & ([\mathbf{Q}_A\mathbf{p}_A\mathbf{e}_A\mathbf{Q}^*_A\mathbf{Q}_B\mathbf{e}_B\mathbf{Q}^*_B] + \\ & [(\mathbf{x}_A - \mathbf{Q}_B\mathbf{p}_B\mathbf{Q}^*_B - \mathbf{x}_B)\mathbf{Q}_A\mathbf{e}_A\mathbf{Q}^*_A\mathbf{Q}_B\mathbf{e}_B\mathbf{Q}^*_B] > 0) \end{aligned} \tag{2.36}$$

This form is not homogeneous, but by rearranging scalar triple products and using the distributivity of rotation over cross product, we obtain

$$\begin{aligned} C^{+}_{e_A,e_B} = & ([\mathbf{Q}_A\mathbf{p}_A\mathbf{e}_A\mathbf{Q}^*_A\mathbf{Q}_B\mathbf{e}_B\mathbf{Q}^*_B] + [\mathbf{Q}_B\mathbf{p}_B\mathbf{e}_B\mathbf{Q}^*_B\mathbf{Q}_A\mathbf{e}_A\mathbf{Q}^*_A] + \\ & [(\mathbf{x}_A - \mathbf{x}_B)\mathbf{Q}_A\mathbf{e}_A\mathbf{Q}^*_A\mathbf{Q}_B\mathbf{e}_B\mathbf{Q}^*_B] > 0) \end{aligned} \tag{2.37}$$

and this equation is independent of the magnitude of either quaternion. Thus we can represent rotations using the vector parts of the quaternions only i.e. $\mathbf{q}_A$ and $\mathbf{q}_B$, and assume the scalar parts have value 1.

For motion planning, the configuration space is now 12-dimensional and configurations are of the form $(\mathbf{x}_A, \mathbf{q}_A, \mathbf{x}_B, \mathbf{q}_B)$. The configuration obstacles in this space are again defined by (2.1) or (2.9), but where A, C^+, etc. constraints have the generalized forms given above. The new constraints are algebraic but their total degree is now five instead of three. Each constraint is linear in the position coordinates of the configuration and quartic in the rotation coordinates.

The collision detection scheme described in the last section can be readily extended to deal with the generalized constraints. Suppose the two objects move uniformly from configuration $(\mathbf{x}_{Ai}, \mathbf{q}_{Ai}, \mathbf{x}_{Bi}, \mathbf{q}_{Bi})$ to configuration $(\mathbf{x}_{Af}, \mathbf{q}_{Af}, \mathbf{x}_{Bf}, \mathbf{q}_{Bf})$. Each coordinate of the objects' configuration will be a linear function of some parameter s, and since the total degree of the constraints is five, each constraint will be a quintic polynomial in the parameter s. Quintics can not be solved explicitly, but numerical methods such as the modified Uspensky algorithm [CL] can be used to find all the zeros of a quintic very rapidly. The zeros of the quintics correspond to true-false transitions of predicates, and these can be merged exactly as described in section 2.4, and the time bounds derived there are still valid.

Thus collision detection for pairs of moving objects has the same asymptotic complexity as collision detection for a moving object and a fixed obstacle. The ratio of running times for the two methods is expected to be small (less than an order of magnitude) although this has not been verified in an implementation. For several moving objects, we find ranges of values of s which lead to collisions between pairs of objects, and then find the union of these ranges for all pairs. There are $O(n^2)$ such pairs to be tested for n moving objects, so the overall complexity is still quadratic.

2.6 Constraints for Jointed Manipulators

In section 2.3 we described the representation of configuration constraints for a polyhedron with six degrees of freedom. In this section we examine robots consisting of polyhedral links with rotary or sliding joints. Our notation mostly follows that of Paul [Pau]. Let $l_1, \ldots, l_n$ be the n links numbered according to their order in the kinematic chain from the base of the robot. We can associate to each link a coordinate frame which is fixed to that link. Transformations between links can be described using 4×4 homogeneous matrices of the following form:

$$A = \begin{pmatrix} U & d \\ 0 & 1 \end{pmatrix} \tag{2.38}$$

where U is a 3×3 unitary matrix representing a rotation, d is a 3×1 vector representing the translation between link coordinate frames. For example, to transform a vector v according to the transformation A, we compute the product

$$\begin{pmatrix} v' \\ 1 \end{pmatrix} = \begin{pmatrix} U & d \\ 0 & 1 \end{pmatrix} \begin{pmatrix} v \\ 1 \end{pmatrix} \tag{2.39}$$

which yields a new vector v' which has been rotated according to U and translated by the vector d. When used to specify the transformation between links of a robot, these matrices are usually called "A" matrices, see Denavit-Hartenberg [DH]. Each A matrix specifies the transformation from the ith link coordinate frame to the $(i-1)^{st}$. For a rotary joint, the transformation matrix has the following form:

$$A = \begin{pmatrix} \cos\theta & -\sin\theta\cos\alpha & \sin\theta\sin\alpha & a\cos\theta \\ \sin\theta & \cos\theta\cos\alpha & -\cos\theta\sin\alpha & a\sin\theta \\ 0 & \sin\alpha & \cos\alpha & d \\ 0 & 0 & 0 & 1 \end{pmatrix} \tag{2.40}$$

where α is the fixed "twist" between successive joint axes, and θ is the (variable) joint angle. a and d are distances between coordinate frame origins in certain directions, and are constant. See Paul's book [Pau] for full description of these parameters. For a sliding joint manipulator, the A matrix has the form

$$A = \begin{pmatrix} \cos\theta & -\sin\theta\cos\alpha & \sin\theta\sin\alpha & 0 \\ \sin\theta & \cos\theta\cos\alpha & -\cos\theta\sin\alpha & 0 \\ 0 & \sin\alpha & \cos\alpha & d \\ 0 & 0 & 0 & 1 \end{pmatrix} \tag{2.41}$$

where d is the variable distance, and θ and α are fixed angles. Recalling that A_i denotes the transformation from the ith link coordinate frame to the $(i-1)^{st}$, we can compute transformations between non-consecutive links by composing A matrices. If we denote the transformation from the jth coordinate frame to the ith (assuming $j > i$) as T_{ij}, then T_{ij} is given by

$$T_{ij} = A_{i+1}A_{i+2}\cdots A_j \tag{2.42}$$

and in particular, T_{0j} is the transformation from the jth link coordinate frame to the base coordinate frame. Constraints on the configuration of the robot may be generated by any two polyhedra lying on any two links of the arm. We treat obstacles in the environment as being part of the 0th link of the manipulator. If we take the disjunction of overlap predicates for all such pairs of polyhedra, we obtain a predicate which is true at a configuration if and only if some link of the robot overlaps some other link or some part of the environment. So it suffices to define the overlap predicate for two polyhedra, one attached to link i and the other attached to link j.

We again make use of the overlap predicates defined in section 2.1. Let A be the polyhedron attached to link i, and B be the polyhedron attached to link j. We compute the positions of vertices, directions of edges etc. in link i's reference frame, so that the vertices, edges and faces of object A are fixed. The features of object B, however, must be transformed by T_{ij}. This transformation has a different form for each type of feature. Vertices, as we have seen, transform according to (2.39). Edge directions, and face normals however, are not displaced by the translational part of the transformation. We can perform a purely rotational transformation by evaluating:

$$\begin{pmatrix} \mathbf{v}' \\ 0 \end{pmatrix} = T_{ij} \begin{pmatrix} \mathbf{v} \\ 0 \end{pmatrix} \tag{2.43}$$

where $\mathbf{v}$ is the face normal or edge direction in link j coordinates, and $\mathbf{v}'$ is the rotated feature in link i coordinates. For face distances c, we must compute the inner product of the rotated face normal with the translational part of the transformation, which is the given by the first three elements of the last column of T_{ij}, that is

$$c' = c + t_{ij}^T T_{ij} \begin{pmatrix} \mathbf{n} \\ 0 \end{pmatrix} \tag{2.44}$$

where t_{ij}^T is the transpose of the last column of T_{ij}, and $\mathbf{n}$ is the untransformed normal of the face.

We substitute transformed expressions for vertices, edges and faces of the polyhedron B and untransformed expressions for polyhedron A into the A and C^+ predicates, (equations (2.3) and (2.7)). This gives predicates which are algebraic in the coefficients of the matrix T_{ij}, and therefore in the coefficients of the matrices A_k for $i < k \leq j$. If the ith joint is a sliding joint, the matrix coefficients of A_i depend linearly on the joint distance. If this joint is a rotary joint, the matrix coefficients are not algebraic in the joint angle, however, there is a simple parametrization of joint angle for which they are algebraic. We let

$$\begin{aligned} u &= \sin\tfrac{\theta}{2} \\ v &= \cos\tfrac{\theta}{2} \end{aligned} \tag{2.45}$$

so that

$$\begin{aligned} \sin\theta &= 2uv \\ \cos\theta &= v^2 - u^2 \end{aligned} \tag{2.46}$$

The reason for choosing this particular representation over any other is that it allows us to express all matrix coefficients as homogeneous polynomials of degree 2 in u and v. That is, the coefficients of the matrix A_i are all either multiples of $\sin\theta$ or $\cos\theta$, or independent of θ. We can substitute for $\sin\theta$ and $\cos\theta$ from (2.46) into A_i, and if we in addition multiply all coefficients that are independent of θ by $u^2 + v^2$, we obtain a matrix all of whose coefficients are homogeneous and of degree 2 in u and v. This means that T_{ij} and the constraint predicates A and C^+ will be homogeneous and of even degree in u and v, which allows us to fix, say $v = 1$ without affecting the value of the predicate, so that the joint angle of the ith joint is parametrized by a single parameter u (u defines a joint angle of $\theta = 2\arctan(u)$ in this case). We can do this for each joint, and so all the constraint predicates are algebraic in a certain parametrization of the joint angles. Thus we can perform collision detection for algebraic paths as described in the last section, or we can use the overlap predicate in the motion planning algorithm described in chapter 4.

Instead of setting $v = 1$, we can perform a substitution for pairs of joints which lowers the algebraic complexity of the constraints. Assuming the constraints are homogeneous in every pair of joint parameters u, v, the parameter space can be thought of as a product of one-dimensional projective spaces P^1, where each space is the set of rays (u, v). The substitution is based on Segre embedding [Mum], which is a method of embedding a product of projective spaces in a single higher-dimensional projective space. Let the joint parameters for the ith joint be denoted u_i, v_i and those for the $(i+1)^{st}$ joint be denoted u_{i+1}, v_{i+1}. Then we define new parameters a, b, c, d as

$$a = u_i u_{i+1} \quad b = u_i v_{i+1} \quad c = v_i u_{i+1} \quad d = v_i v_{i+1} \tag{2.47}$$

Since this gives us four parameters, when we really have only two degrees of freedom, we need to add "commutation relations" which in this case consists of the single polynomial

$$ad = bc \tag{2.48}$$

Then those constraints that previously had degree two in u_i and v_i and degree two in u_{i+1} and v_{i+1}, i.e. total degree four, now have total degree two in a, b, c, d. Furthermore, the constraint equations should be homogeneous in a, b, c, d so that we are now in the projective space P^3. We can again use the projection technique which we used in the section on quaternions, by setting say $d = 1$. This gives us a hyperplane which covers most (all but a lower dimensional subset) of P^3 when projected onto it. To be strictly correct, in order to get all solutions in P^3, we should use the "four cubes" representation described in section 2.3.1.

Chapter 3

Elimination Theory

Elimination theory evolved during the period of constructive methods in algebra. It is the study of conditions for the existence of solutions of systems of polynomial equations. Its main results were known at least a century ago [Mer], [Hur] and [Wae] and still appear in modern treatments of algebraic geometry, although usually in non-constructive form. A central result which seems to have disappeared from modern treatments is the construction of a single resultant polynomial for a system of n homogeneous polynomials in n unknowns. This resultant is zero if and only if the system has a non-trivial solution. The term resultant has since come to be applied only to the resultant of two polynomials, also known as the Sylvester resultant. For this reason we will use the term *multivariate resultant* to distinguish the more general case. A familiar example of such a condition is the determinant of a system of linear equations. The multivariate resultant is a generalization of both the determinant and the Sylvester resultant, and as we shall see below it includes both as special cases.

The early work demonstrated that the multivariate resultant not only gives a necessary and sufficient condition for solvability of the system, but that it is generically irreducible, so that it is the smallest such condition. In spite of its evident utility, it does not appear to have been used previously as a computational tool. In this chapter we give an algorithm for the computation of the multivariate resultant which runs in single exponential time, with exponent equal to the number of variables.[1] The algorithm also runs in polynomial time in parallel, with an exponential number of processors.

We should point out here that the matrices used in the computation of the multivariate resultant are related to those used in Lazard's method [Laz81], and its generalization by Grigoryev, Chistov and Vorobjov, [CG], [Gri], [GV]. These algorithms form a central component of the single-exponential time decision algorithms for existential algebraic theories which we described in the introduction.[2] These methods in fact compute inertia forms (defined later in this chapter) which are multiples of the resultant, and therefore give only necessary conditions for solvability. They also do not appear to be readily parallelizable. On the other hand, the present algorithm is highly uniform, and we give a version of it which involves neither division nor branching, and can be computed in polynomial time in parallel.

The multivariate resultant algorithm is the central computational tool in the roadmap algorithm described in the next chapter. The other applications of multivariate resultants extend well beyond the scope of this thesis. In this chapter we describe only two, the first being an improved "gap" theorem, and the second being a fast parallel method for solving systems of polynomial equations.

3.1 The Multivariate Resultant

For n homogeneous polynomials $f_1, \ldots, f_n$ of degrees $d_1, \ldots, d_n$ in the n variables $x_1, \ldots, x_n$, there exists a single resultant R, which is a homogeneous polynomial in the coefficients of the f_i, [Wae]. The vanishing of this resultant is necessary and sufficient for the polynomials to have a non-trivial simultaneous solution (a solution distinct from 0) in the algebraic closure of the base field. This resultant has degree $d_1, d_2, \ldots, d_{i-1}d_{i+1}, \ldots, d_n$ in the

[1] At the time when this thesis was written, the author was unaware of earlier expressions for the multivariate resultant due to Cayley [Sal] and Macaulay [Mac]. However, the author's new formulation can be implemented without division, and does not fail on certain systems of polynomials, as did the earlier methods. Another modern algorithm based on the resultant is described in [Ren]

[2] The bounds for the existential theory of the reals (and for the theory of the complexes, which is strictly easier) have since been improved to polynomial space, or polynomial parallel time, in a recent paper of the author's [Ca88a]

coefficients of f_i. While R is not readily expressible as the determinant of a single matrix, it can be defined as the greatest common divisor of the determinants of n matrices $A^1, \ldots, A^n$. As we shall see below, it is also possible to define the resultant as a rational function of certain subdeterminants of the A^k thereby avoiding the GCD computation. In the special case $n = 2$, the determinants of A_1 and A_2 are identical, and both are equal to the Sylvester resultant of f_1 and f_2.

The matrices A^k are constructed as follows. The description follows that given in [Wae], and the formulation is due originally to Hurwirtz [Hur]. First we define the degree d of monomials that index the columns of A^k:

$$d = 1 + \sum_{i=1,\ldots,n} (d_i - 1) \tag{3.1}$$

Let $\alpha \in Z^n$ represent the exponents of a monomial in $x_1, \ldots, x_n$, i.e. if $\alpha = (\alpha_1, \alpha_2, \ldots, \alpha_n)$ we use the notation x^α for the monomial

$$x^\alpha = x_1^{\alpha_1} x_2^{\alpha_2} \ldots, x_n^{\alpha_n}$$

Then the set of monomials of degree d, denoted X^d is

$$X^d = \{x^\alpha \mid \alpha_1 + \alpha_2 + \cdots + \alpha_n = d\} \tag{3.2}$$

and we observe that the cardinality of X^d is

$$N = |X^d| = \binom{d+n-1}{d} \tag{3.3}$$

We now partition X^d into n subsets X_i^d as follows: X_i^d is the set of monomials which are multiples of $x_i^{d_i}$, but which are not multiples of $x_j^{d_j}$ for any $j < i$, so that

$$\begin{aligned}
X_1^d &= \{x^\alpha \in X^d \mid \alpha_1 \geq d_1\} \\
X_2^d &= \{x^\alpha \in X^d \mid \alpha_2 \geq d_2 \text{ and } \alpha_1 < d_1\} \\
&\vdots \qquad \vdots \\
X_n^d &= \{x^\alpha \in X^d \mid \alpha_n \geq d_n \text{ and } \alpha_i < d_i, \text{ for } i = 1, \ldots, n-1\}
\end{aligned} \tag{3.4}$$

and it is readily shown that the X_i^d are disjoint, and that every element of X^d is contained in exactly one of them. Now for each X_i^d we define a set of polynomials F_i as

$$F_i = \frac{X_i^d}{x_i^{d_i}} f_i \tag{3.5}$$

that is, $F_i = \{x^\beta f_i \mid x^\beta x_i^{d_i} \in X_i^d\}$. Then the polynomials in F_i, for $i = 1, \ldots, n$, have degree d. The union F of the F_i is a collection of N polynomials of degree d. Since there are N monomials of degree d, we can construct a square matrix whose columns correspond to the N monomials in X^d, and each row of which contains the coefficients of those monomials in some polynomial in F. Specifically, we choose any ordering of the N elements of X^d, and then construct an $N \times N$ matrix A^n where A^n_{ij} is the coefficient of the ith monomial of X^d in the jth polynomial of F. The determinant of A^n has degree $d_1 d_2, \ldots, d_{n-1}$ in the coefficients of f_n, because the number of monomials in X_n^d is $d_1 d_2, \ldots, d_{n-1}$.

The A^k for $k = 1, \ldots, n-1$ are defined similarly, but the sets of monomials X_i^d are different. In the definition (3.4) each X_i^d is the set of elements of X^d which are multiples of $x_i^{d_i}$, which are not contained in previous X_i^d's. For A^k, we reorder the X_i^d so that X_k^d appears last in the definition, so that it is the set of all multiples of $x_k^{d_k}$ which are not multiples of $x_i^{d_i}$ for any $i \neq k$. Thus X_k^d has $d_1, \ldots, d_{k-1} d_{k+1}, \ldots, d_n$ elements, and this number is also the degree of $\det A^k$ in the coefficients of f_k.

From now on, we will let a^k denote the determinant of A^k. If the polynomials f_i have indeterminate (symbolic) coefficients, the a^k are known as inertia forms [Hur]. They are characterized by the following property:

$$x_j^p a^k = 0(f_1, \ldots, f_n) \tag{3.6}$$

holds for some p and some (in fact every) j.

Now the resultant of $f_1, \ldots, f_n$ is the greatest common divisor of the a^k *as polynomials in the indeterminate coefficients of* $f_1, \ldots, f_n$ (see [Wae], section 82). This is a crucial point. We are interested in the value of the resultant for a certain specialization of the coefficients. By a specialization of the coefficients, we mean the introduction of polynomial relations on the indeterminate coefficients, e.g. $u_1 = 5$ or $u_1 = 2u_2$. This only succeeds up to a point. Specialization commutes with addition and multiplication and therefore with determinant computation. Thus we get the same result for each a^k by evaluating a^k formally and substituting for the coefficients as we do by substituting first and evaluating the determinant numerically.

However, specialization does not commute with the greatest common divisor step. For example, consider the greatest common divisor of ax and bx, where a and b are indeterminates. Suppose that a and b are both specialized to zero. If we compute the GCD before specialization, we get x, but both terms vanish if we specialize before the GCD step. For the Sylvester resultant, the GCD step is unnecessary, because the two a^k are identical. Similarly, if all the polynomials are linear, then all the a^k are identical, and equal to the determinant of the system. For the general case, the a^k are different, and the GCD step is non-trivial. One can require that the specialization of the coefficients be sufficiently "generic", since the GCD step succeeds for almost all specializations of the coefficients. However, this is an unnecessary restriction. At the other extreme, it is possible to compute the determinants symbolically, and then compute the GCD of the a^k as polynomials in *all* the symbolic coefficients of the f_i. However, this is a massive task, and generates terms of size double exponential in d.

3.1.1 Partial Coefficient Specialization

In the method described here, we instead specialize all but n of the coefficients (Strictly speaking, rather than leaving some coefficients indeterminate, we add an indeterminate offset to those coefficients). This ensures that the GCD of the a^k is the resultant. Furthermore, it allows the resultant to be recovered without a true GCD step, in fact without division over the base field. The resultant is computed from the a^k by a sequence of $2n$ polynomials divisions involving only *monic* polynomials. A monic polynomial is by definition a polynomial whose leading coefficient is one. Division of monic polynomials does not involve division of their coefficients.

We define a new system of polynomials $\hat{f}_i$ from the f_i as follows:

$$\hat{f}_i(x_1, \ldots, x_n) = f_i(x_1, \ldots, x_n) + u_i x_i^{d_i} \tag{3.7}$$

where the u_i are indeterminates. Notice that the u_i appear on the leading diagonal of every matrix A^k. In fact u_i occupies the leading diagonal position in the columns indexed by monomials in X_i^d. The product of the diagonal u_i's appears as the leading term of a^k, treating a^k as a polynomial in $u_1, \ldots, u_n$, and its coefficient is 1. It follows that a^k has degree $d_1, \ldots, d_{k-1}d_{k+1}, \ldots, d_n$ in u_k. But the resultant R also has degree $d_1, \ldots, d_{i-1}d_{i+1} \ldots, d_n$ in u_i for all i. In fact the resultant (before specialization) has a single term of maximal degree in all the u_i, and this term has coefficient 1 (see [Wae]). Since the resultant divides all the a^k, we can write

$$a^k(u_1, \ldots, u_n) = b^k(u_1, \ldots, u_{k-1}, u_{k+1}, \ldots, u_n)R(u_1, \ldots, u_n) \qquad (3.8)$$

where b^k is a polynomial independent of u_k because R and a^k have the same degree in u_k. Knowing the a^k, and knowing that the leading coefficients of both the a^k and R are 1, we can actually solve for all the coefficients of R (and b^k if desired). First we need:

Definition A multivariate polynomial $p(x_1, \ldots, x_2)$ of degree d_i in x_i is said to be *rectangular* if its leading term is of maximal degree in all the x_i, i.e. the leading term is of the form $cx_1^{d_1}x_2^{d_2} \ldots, x_n^{d_n}$. The product or quotient of two rectangular polynomials is also rectangular.

In particular, $R(u_1, \ldots, u_n)$ is rectangular and monic, i.e. its leading coefficient is one. All the $a^k(u_1, \ldots, u_n)$ are rectangular and monic, so the $b^k(u_1, \ldots, u_{k-1}, u_{k+1}, \ldots, u_n)$ are rectangular and monic. To compute the resultant we can use:

The Basic Recurrence

Let $R_j(u_1, \ldots, u_j)$ be the leading coefficient of R considered as a polynomial in $u_{j+1}, \ldots, u_n$. Since R is rectangular, it follows that the leading coefficient of R_j equals the leading coefficient of R, so that R_j is monic. Let $b_j^k(u_1, \ldots, u_j)$ be the leading coefficient of b^k, considered as a polynomial in $u_{j+1}, \ldots, u_n$. Then b_j^k is also monic. Finally, let $a_j^k(u_1, \ldots, u_j)$ be the leading coefficient of a^k considered as a polynomial in $u_{j+1}, \ldots, u_n$. Now the polynomial identity (3.8) implies a corresponding identity on the leading coefficients of the polynomials, so the following relationships hold:

$$\begin{aligned} R_{k-1}b_{k-1}^{k} &= a_{k-1}^{k} \qquad \text{for } k = 1, \ldots, n \\ R_k b_{k-1}^{k} &= a_{k}^{k} \qquad \text{for } k = 1, \ldots, n \end{aligned} \tag{3.9}$$

Where the second expression exploits the fact that $b_k^k = b_{k-1}^k$ since b^k is independent of u_k. So if we know R_{k-1}, we can compute R_k by first dividing a_{k-1}^k by R_{k-1} to get b_{k-1}^k, and then dividing b_{k-1}^k into a_k^k. Since we know $R_0 = 1$, after $2n$ divisions we obtain $R_n = R$. Notice that all of the polynomials in the division process are monic, and division by a monic polynomial involves only multiplication and subtraction steps on the coefficients. Thus, the computation of the GCD using the recurrence in (3.9) commutes with specialization. Having finally obtained R as a polynomial in $u_1, \ldots, u_n$, we specialize $u_1, \ldots, u_n$ to zero, yielding the desired resultant.

3.1.2 Method 1

The recurrence relations just described allow the resultant to be computed for any specialization of the coefficients of the f_i, but involve computation of determinants which are high degree polynomials in $u_1, \ldots, u_n$, rather than elements of the base field. In fact we are only interested in the constant term of $R(u_1, \ldots, u_n)$, since the original system corresponds to the specialization $u_i = 0$ for all i. The polynomial identities in (3.9) imply corresponding identities on the *constant terms* of these polynomials.

So if $R_{k,0}$ denotes the constant term of R_k etc., we have

$$R_{k,0} = \frac{a_{k,0}^{k} R_{k-1,0}}{a_{k-1,0}^{k}} \tag{3.10}$$

where once again $R_{0,0} = 1$. Each $a_{j,0}^k$ is a subdeterminant of a^k with u_i specialized to zero, so that no manipulation of polynomials in u_i is necessary. This method involves only slightly more computation than evaluation of the determinant of A^n *for the original polynomial system.*

The disadvantage of this method is that it is not guaranteed to succeed, since some of the $a_{j,0}^k$ may vanish, even if the resultant does not. One possible solution is to evaluate (3.9) by evaluating *only a term of lowest degree* in each R_k and b^k_{k-1}, since computing the term of lowest degree in a polynomial quotient only requires knowledge of the term of lowest degree in the divisor. However, in the worst case, this may involve evaluation of all the terms in the a^k. Furthermore, for large n, the intermediate $a^k_{j,0}$ will be much larger than the resultant itself, so that their computation seems wasteful. Instead we describe a second method for which the a^k_j cannot vanish, so that branching is unnecessary. In the latter method, division over the base field is never necessary, and intermediate terms larger than the resultant are never generated.

3.1.3 Method 2

The recurrence relation in (3.9) is valid for any specialization of the $u_1, \ldots, u_n$. The problem with the recurrence as it stands is that it involves polynomials of quite high degree in n variables. If we add the specialization $u_1 = u_2 = \cdots = u_n$, then all polynomials become univariate in the single variable u_1. The effect of this specialization is to collapse all the terms of a given degree (by summation) into the coefficient of u_1 of that degree. From the definition of rectangularity, it should be clear that the polynomials remain monic after this specialization. Thus the method now involves only division of monic polynomials in the single variable u_1. We denote the polynomials after this specialization by $\bar{R}$, $\bar{a}^k_j$, $\bar{b}^k_j$, so that $\bar{R}(u_1) = R(u_1, u_1, \ldots, u_1)$ and $\bar{a}^k_j(u_1) = a^k_j(u_1, u_1, \ldots, u_1)$, etc. Then we have the recursion

$$
\begin{aligned}
\bar{b}^k_{k-1}(u_1) &= \frac{\bar{a}^k_{k-1}(u_1)}{\bar{R}_{k-1}(u_1)} \qquad \text{for } k = 1, \ldots, n \\
\bar{R}_k(u_1) &= \frac{\bar{a}^k_k(u_1)}{\bar{b}^k_{k-1}(u_1)} \qquad \text{for } k = 1, \ldots, n
\end{aligned}
\tag{3.11}
$$

Recall from the properties of the A^k, that the indeterminates u_i lie along the leading diagonal. Since a^k_j is the leading coefficient (term of highest degree) with respect to the variables $u_{j+1}, \ldots, u_n$, a^k_j is the determinant of the submatrix of A^k obtained by deleting all those rows and columns that contain any of the variables $u_{j+1}, \ldots, u_n$.

For notational expendiency, we define A^k_j as the submatrix of A^k whose determinant is a^k_j. In fact we can define A^k_j without making use of any of the introduced indeterminates u_i. Let A_j^k be the submatrix of A^k obtained by deleting all rows and columns that contain the coefficient of $x_i^{d_i}$ in f_i for $i > j$. This definition simply follows from the fact that these are the rows and columns that *would have* contained the introduced variables $u_{j+1}, \ldots, u_n$.

Then the constant coefficient $a_{j,0}^k$ is simply the determinant of A_j^k with no indeterminates introduced. For $\bar{a}_j^k(u_1)$, we notice that the indeterminate $u_1 = \cdots = u_j$ appears on the leading diagonal of A_j^k, so that $\bar{a}_j^k(u_1)$ is the characteristic polynomial of A_j^k. For computation in parallel, the known fast methods for computing the determinant [Csa] in fact compute the coefficients of the characteristic polynomial as a side effect, so method 2 is as fast as method 1 using current parallel algorithms.

Notice also that division by monic polynomials does not require division in the base field. The coefficients of the quotient can be computed as principal minors of a certain matrix. Let $a(s)$ and $b(s)$ be two polynomials of degree m, n respectively, $m > n$, and where b is monic. Then to find the quotient of a and b we form the following matrix:

$$Q = \begin{pmatrix} a_m & a_{m-1} & a_{m-2} & \cdots & a_{n+1} & a_n & \cdots & a_1 & a_0 \\ -1 & -b_{n-1} & -b_{n-2} & \cdots & -b_{2n-m+1} & -b_{2n-m} & \cdots & 0 & 0 \\ 0 & -1 & -b_{n-1} & \cdots & -b_{2n-m+2} & -b_{2n-m+1} & \cdots & 0 & 0 \\ \vdots & \vdots & \vdots & & \vdots & \vdots & & \vdots & \vdots \\ 0 & 0 & 0 & \cdots & -1 & -b_{n-1} & \cdots & -b_0 & 0 \end{pmatrix} \tag{3.12}$$

and this matrix has $m - n + 1$ rows. The left-most $m - n + 1$ columns form a square matrix, and it is readily seen that the principal subdeterminants are the coefficients of the quotient a/b. That is, the coefficient of s^i in a/b is the determinant of the submatrix containing the first $m - n - i + 1$ rows and columns of Q. This relationship shows also that the quotient depends only on the leading $m - n + 1$ coefficients of both polynomials, so that it is not necessary to compute the lower order terms. The leading r terms of a product of polynomials also depends only on the leading r terms of each of the polynomials. So in the recurrence (3.11), we need only evaluate a number of terms in each of the polynomials equal to the degree of the resultant polynomial $\bar{R}_n(u_1)$.

It turns out that for characteristic polynomials, which have leading coefficient one, the coefficient size increases linearly as degree decreases. (see section **3.3** on polynomial height for a precise characterization of this). The higher order terms are the least expensive to compute. If the characteristic polynomial of a matrix A is of the form $C(s) = s^n + c_1 s^{(n-1)} + \cdots + c_n$, then the coefficients c_i can be found from the following equation, sometimes called the Newton expansion:

$$\begin{pmatrix} 1 & & & & & \\ s_1 & 2 & & & & \\ s_2 & s_1 & 3 & & & \\ s_3 & s_2 & s_1 & 4 & & \\ \vdots & \vdots & \vdots & \vdots & \ddots & \\ s_{n-1} & s_{n-2} & s_{n-3} & s_{n-4} & \cdots & n \end{pmatrix} \begin{pmatrix} c_1 \\ c_2 \\ c_3 \\ c_4 \\ \vdots \\ c_n \end{pmatrix} = - \begin{pmatrix} s_1 \\ s_2 \\ s_3 \\ s_4 \\ \vdots \\ s_n \end{pmatrix} \tag{3.13}$$

where s_j is the trace of the jth power of the matrix A. Notice that we can compute the first M characteristic polynomial coefficients by solving an $M \times M$ system, using the upper left $M \times M$ submatrix of (3.13). So instead of solving $N \times N$ systems to find all the coefficients of the characteristic polynomial of the A^k, we need only solve a series of $M \times M$ systems, where M is the degree of the resultant $\overline{R}_n(u_1)$ in u_1. By the argument of the previous paragraph, only this number of coefficients is required for computation of polynomial quotients.

For a system of n polynomials of degree d, the characteristic polynomial of an A^k has degree approximately $(de)^n$ in u_1 where e is the base of natural logarithms, while the resultant has degree $nd^{(n-1)}$. For certain problems, such as equation solving, the saving caused by evaluating only leading coefficients of the a_j^k dominates the extra effort involved in doing division of polynomials in u_1 rather than elements of the base field (which may themselves be multivariate polynomials).

Since method 2 involves only computation of determinants of matrices of single exponential size, and since monic polynomial division can also be performed using determinant computations, Csanky's $\log^2$ parallel time algorithm [Csa] implies that the resultant can be computed in polynomial parallel time (in fact in PSPACE).

An Example We compute the resultant of a system of 2 quadrics and a linear polynomial in three variables x, y, z. The polynomials are:

$$
\begin{aligned}
f_1 &= a_{xx}x^2 + a_{xy}xy + a_{xz}xz + a_{yy}y^2 + a_{yz}yz + a_{zz}z^2 \\
f_2 &= b_{xx}x^2 + b_{xy}xy + b_{xz}xz + b_{yy}y^2 + b_{yz}yz + b_{zz}z^2 \\
f_3 &= c_x x + c_y y + c_z z
\end{aligned}
\tag{3.14}
$$

Then to construct the matrix A^1, we first notice that $d = \sum(d_i - 1) + 1 = 3$. For the purposes of the construction, we order the polynomials as f_2, f_3, f_1 so that f_1 is last in the list. Then we split the monomials of degree 3 into three groups, each of which contains all the remaining monomials which are divisible by $x_i^{d_i}$. In this case, the monomials in each group are divisible by y^2, z and x^2 respectively:

$$
\begin{aligned}
X_2^3 &= \{y^3, y^2z, y^2x\} \\
X_3^3 &= \{yz^2, yzx, z^3, z^2x, zx^2\} \\
X_1^3 &= \{yx^2, x^3\}
\end{aligned}
\tag{3.15}
$$

We use these monomials to index the rows and columns of A^1. Each row of A^1 is filled with the coefficients of $\frac{x^\alpha}{x_i^{d_i}} f_i$ where the monomial x^α is the row index, and is contained in X_i^3.

$$A^1 = \begin{array}{c} \\ y^3 \\ y^2z \\ y^2x \\ yz^2 \\ yzx \\ z^3 \\ z^2x \\ zx^2 \\ yx^2 \\ x^3 \end{array} \begin{array}{c} \begin{array}{cccccccccc} y^3 & y^2z & y^2x & yz^2 & yzx & z^3 & z^2x & zx^2 & yx^2 & x^3 \end{array} \\ \left(\begin{array}{cccccccccc} b_{yy} & b_{yz} & b_{xy} & b_{zz} & b_{xz} & 0 & 0 & 0 & b_{xx} & 0 \\ 0 & b_{yy} & 0 & b_{yz} & b_{xy} & b_{zz} & b_{xz} & b_{xx} & 0 & 0 \\ 0 & 0 & b_{yy} & 0 & b_{yz} & 0 & b_{zz} & b_{xz} & b_{xy} & b_{xx} \\ 0 & c_y & 0 & c_z & c_x & 0 & 0 & 0 & 0 & 0 \\ 0 & 0 & c_y & 0 & c_z & 0 & 0 & 0 & c_x & 0 \\ 0 & 0 & 0 & c_y & 0 & c_z & c_x & 0 & 0 & 0 \\ 0 & 0 & 0 & 0 & c_y & 0 & c_z & c_x & 0 & 0 \\ 0 & 0 & 0 & 0 & 0 & 0 & 0 & c_z & c_y & c_x \\ a_{yy} & a_{yz} & a_{xy} & a_{zz} & a_{xz} & 0 & 0 & 0 & a_{xx} & 0 \\ 0 & 0 & a_{yy} & 0 & a_{yz} & 0 & a_{zz} & a_{xz} & a_{xy} & a_{xx} \end{array} \right) \end{array} \tag{3.16}$$

Notice that the coefficients of $x_i^{d_i}$ in f_i, namely b_{yy}, c_z and a_{xx} appear along the leading diagonal. A^2 is similar to A^1 and can be formed by simply interchanging x and y.

$$A^2 = \begin{array}{c} \\ x^3 \\ x^2z \\ x^2y \\ xz^2 \\ xzy \\ z^3 \\ z^2y \\ zy^2 \\ xy^2 \\ y^3 \end{array} \begin{array}{c} \begin{array}{cccccccccc} x^3 & x^2z & x^2y & xz^2 & xzy & z^3 & z^2y & zy^2 & xy^2 & y^3 \end{array} \\ \left(\begin{array}{cccccccccc} a_{xx} & a_{xz} & a_{xy} & a_{zz} & a_{yz} & 0 & 0 & 0 & a_{yy} & 0 \\ 0 & a_{xx} & 0 & a_{xz} & a_{xy} & a_{zz} & a_{yz} & a_{yy} & 0 & 0 \\ 0 & 0 & a_{xx} & 0 & a_{xz} & 0 & a_{zz} & a_{yz} & a_{xy} & a_{yy} \\ 0 & c_x & 0 & c_z & c_y & 0 & 0 & 0 & 0 & 0 \\ 0 & 0 & c_x & 0 & c_z & 0 & 0 & 0 & c_y & 0 \\ 0 & 0 & 0 & c_x & 0 & c_z & c_y & 0 & 0 & 0 \\ 0 & 0 & 0 & 0 & c_x & 0 & c_z & c_y & 0 & 0 \\ 0 & 0 & 0 & 0 & 0 & 0 & 0 & c_z & c_x & c_y \\ b_{xx} & b_{xz} & b_{xy} & b_{zz} & b_{yz} & 0 & 0 & 0 & b_{yy} & 0 \\ 0 & 0 & b_{xx} & 0 & b_{xz} & 0 & b_{zz} & b_{yz} & b_{xy} & b_{yy} \end{array} \right) \end{array} \tag{3.17}$$

Finally, for A^3, we use the order f_1, f_2, f_3, and the monomials:

$$\begin{aligned} X_1^3 &= \{x^3, x^2y, x^2z\} \\ X_2^3 &= \{xy^2, y^3, y^2z\} \\ X_3^3 &= \{xyz, xz^2, yz^2, z^3\} \end{aligned} \tag{3.18}$$

and the matrix itself is

$$A^3 = \begin{array}{c} \\ x^3 \\ x^2y \\ x^2z \\ xy^2 \\ y^3 \\ y^2z \\ xyz \\ xz^2 \\ yz^2 \\ z^3 \end{array} \begin{array}{c} \begin{array}{cccccccccc} x^3 & x^2y & x^2z & xy^2 & y^3 & y^2z & xyz & xz^2 & yz^2 & z^3 \end{array} \\ \left(\begin{array}{cccccccccc} a_{xx} & a_{xy} & a_{xz} & a_{yy} & 0 & 0 & a_{yz} & a_{zz} & 0 & 0 \\ 0 & a_{xx} & 0 & a_{xy} & a_{yy} & a_{yz} & a_{xz} & 0 & a_{zz} & 0 \\ 0 & 0 & a_{xx} & 0 & 0 & a_{yy} & a_{xy} & a_{xz} & a_{yz} & a_{zz} \\ b_{xx} & b_{xy} & b_{xz} & b_{yy} & 0 & 0 & b_{yz} & b_{zz} & 0 & 0 \\ 0 & b_{xx} & 0 & b_{xy} & b_{yy} & b_{yz} & b_{xz} & 0 & b_{zz} & 0 \\ 0 & 0 & b_{xx} & 0 & 0 & b_{yy} & b_{xy} & b_{xz} & b_{yz} & b_{zz} \\ 0 & c_x & 0 & c_y & 0 & 0 & c_z & 0 & 0 & 0 \\ 0 & 0 & c_x & 0 & 0 & 0 & c_y & c_z & 0 & 0 \\ 0 & 0 & 0 & 0 & 0 & c_y & c_x & 0 & c_z & 0 \\ 0 & 0 & 0 & 0 & 0 & 0 & 0 & c_x & c_y & c_z \end{array}\right) \end{array} \tag{3.19}$$

Since the matrices all have symbolic coefficients, none of the intermediate terms will vanish, so we can use method 1. Now to start the recurrence we have $a_{0,0}^1 = 1$, and $R_{0,0} = 1$. $a_{1,0}^1$ is the determinant of the matrix A_1^1 which is obtained by deleting all the rows and columns of A^1 which contain the coefficient of $x_i^{d_i}$, for $i > 1$, that is, all the rows and columns containing b_{yy} or c_z. Hence

$$A_1^1 = \begin{pmatrix} a_{xx} & 0 \\ a_{xy} & a_{xx} \end{pmatrix} \tag{3.20}$$

so that $a_{1,0}^1 = a_{xx}^2$. Dividing by $b_{0,0}^1 = 1$ we obtain $R_{1,0} = a_{xx}^2$. Next we need $a_{1,0}^2$ and $a_{2,0}^2$ which are the determinants of the matrices:

$$A_1^2 = \begin{pmatrix} a_{xx} & a_{xz} & a_{xy} \\ 0 & a_{xx} & 0 \\ 0 & 0 & a_{xx} \end{pmatrix} \tag{3.21}$$

$$A_2^2 = \begin{pmatrix} a_{xx} & a_{xz} & a_{xy} & a_{yy} & 0 \\ 0 & a_{xx} & 0 & 0 & 0 \\ 0 & 0 & a_{xx} & a_{xy} & a_{yy} \\ b_{xx} & b_{xz} & b_{xy} & b_{yy} & 0 \\ 0 & 0 & b_{xx} & b_{xy} & b_{yy} \end{pmatrix} \tag{3.22}$$

so that $a_{1,0}^2 = a_{xx}^3$, $b_{1,0}^2 = a_{1,0}^2/R_{1,0} = a_{xx}$, and

$$\begin{aligned} a_{2,0}^2 &= a_{xx}(a_{xx}^2 b_{yy}^2 - a_{xx}a_{xy}b_{xy}b_{yy} - 2a_{xx}a_{yy}b_{xx}b_{yy} + \\ & a_{xx}a_{yy}b_{xy}^2 + a_{xy}^2 b_{xx}b_{yy} - a_{xy}a_{yy}b_{xx}b_{xy} + a_{yy}^2 b_{xx}^2) \end{aligned} \tag{3.23}$$

and since $R_{2,0} = a_{2,0}^2/b_{1,0}^2$, we have

$$\begin{aligned} R_{2,0} &= a_{xx}^2 b_{yy}^2 - a_{xx}a_{xy}b_{xy}b_{yy} - 2a_{xx}a_{yy}b_{xx}b_{yy} + \\ & a_{xx}a_{yy}b_{xy}^2 + a_{xy}^2 b_{xx}b_{yy} - a_{xy}a_{yy}b_{xx}b_{xy} + a_{yy}^2 b_{xx}^2 \end{aligned} \tag{3.24}$$

The next step is the computation of $a_{2,0}^3$ which is the determinant of

$$A_2^3 = \begin{pmatrix} a_{xx} & a_{xy} & a_{xz} & a_{yy} & 0 & 0 \\ 0 & a_{xx} & 0 & a_{xy} & a_{yy} & a_{yz} \\ 0 & 0 & a_{xx} & 0 & 0 & a_{yy} \\ b_{xx} & b_{xy} & b_{xz} & b_{yy} & 0 & 0 \\ 0 & b_{xx} & 0 & b_{xy} & b_{yy} & b_{yz} \\ 0 & 0 & b_{xx} & 0 & 0 & b_{yy} \end{pmatrix} \tag{3.25}$$

which can be factored as

$$\begin{aligned} a_{2,0}^3 &= (a_{xx}b_{yy} - a_{yy}b_{xx})(a_{xx}^2 b_{yy}^2 - a_{xx}a_{xy}b_{xy}b_{yy} - 2a_{xx}a_{yy}b_{xx}b_{yy} + \\ & a_{xx}a_{yy}b_{xy}^2 + a_{xy}^2 b_{xx}b_{yy} - a_{xy}a_{yy}b_{xx}b_{xy} + a_{yy}^2 b_{xx}^2) \end{aligned} \tag{3.26}$$

so that $b_{2,0}^3 = a_{2,0}^3/R_{2,0} = a_{xx}b_{yy} - a_{yy}b_{xx}$. To complete the computation, we compute the determinant of $A_3^3 = A^3$ and divide it by $b_{2,0}^3$ to yield the resultant. Unfortunately, both the determinant and the quotient are too large to be given in this text. However, they have both been computed using Macsyma and the computation time for this problem was about a minute on a Symbolics 3600 Lisp Machine.

3.2 The u-Resultant

We can use the multivariate resultant defined above to find the solutions of a system of polynomials when the number of solutions is finite.[3] Let $f_1, \ldots, f_{n-1}$ be $n-1$ homogeneous polynomials in the unknowns $x_1, \ldots, x_n$. We introduce the linear polynomial with indeterminate coefficients:

$$l = v_1 x_1 + v_2 x_2 + \cdots + v_n x_n \tag{3.27}$$

If we now compute the resultant of $f_1, \ldots, f_{n-1}, l$ in the variables $x_1, \ldots, x_n$, we obtain a resultant $R(v_1, \ldots, v_n)$ which is a homogeneous polynomial whose degree is the product of the degrees of the f_i. This polynomial is called the u-resultant of $f_1, \ldots, f_{n-1}$ [Wae]. If the solution set of $f_1 = 0, \ldots, f_{n-1} = 0$ is finite, then the system $f_1 = 0, \ldots, f_{n-1} = 0, l = 0$ has no solution for almost all linear forms l. Thus the resultant must be non-vanishing. It is easily shown that the u-resultant factors completely over the complex numbers into linear factors of the form:

$$(s_1^{(i)} v_1 + s_2^{(i)} v_2 + \cdots + s_n^{(i)} v_n) \tag{3.28}$$

and that for each factor, $x_1 = s_1^{(i)}$, $x_2 = s_2^{(i)}, \ldots, x_n = s_n^{(i)}$ is a solution of the original system.

Unfortunately, if the system $f_1 = 0, \ldots, f_{n-1} = 0$ has an infinite number of solutions, the u-resultant polynomial vanishes identically. However, it is still possible to recover isolated solution points using the *generalized characteristic polynomial* of the system, instead of the resultant. This approach is described in [Ca88b].

3.3 Degree bounds and Gap Theorems

For the remaining algebraic algorithms, we will be making use of finite precision binary numbers to approximate algebraic numbers. Although this only allows us to evaluate expressions approximately, if we are careful, we can

[3] Another single exponential time algorithm for solving systems of equations was recently given by Renegar [Ren], which is based on the u-resultant and on Macaulay's expression for the multivariate resultant [Mac]

still determine the values of certain expressions exactly using "Gap" theorems. Specifically, these theorems state that for any collection of integral polynomials of a certain size (bounded number of bits in each coefficient and bounded degree) there is a non-zero minimum spacing between any two distinct roots of the polynomials in the collection. We may speak about the size of an algebraic number, this size being the size of the minimal polynomial which has this number as a root.

Thus if the binary approximations to two algebraic numbers of a certain size are sufficiently close (their difference is smaller than the gap), then the two algebraic numbers must be identical. In particular, if the binary approximation to an algebraic number of a certain size is sufficiently small, then the algebraic number must be zero.

It turns out that manipulation of these binary approximations is much more efficient than the manipulation of "exact" algebraic numbers as described by Schwartz and Sharir [SSh]. Roughly speaking, the approximate computations can be done in exponential time, while the exact arithmetic representation grows doubly exponentially with the number of arithmetic operations.

Definition The *weight* of an integral multivariate polynomial f is the sum of the magnitudes of the coefficients of f. We denote the weight of f as $w(f)$, and it is readily shown that $w(f+g) \leq w(f)+w(g)$ and $w(fg) \leq w(f)w(g)$.

Weight is a useful measure for most types of polynomial, but we will also be manipulating characteristic polynomials quite a lot and these have a special structure which makes weight a very poor measure of size. Instead we use:

Definition. The *height* of an integral monic polynomial $f(s)$ of degree m is the smallest h such that $w(f_{m-i}) \leq h^i$, where f_j is the coefficient of s^j in f. We denote the height of f as $h(f)$.

It is easy to show that the characteristic polynomial of an $N \times N$ matrix has height Nw, where w is the maximum weight of any element of the matrix. The utility of height as a measure of polynomial size is emphasized by the following result:

Lemma 3.3.1 *Let $a(s)$, $b(s)$ be monic polynomials. Then $h(ab) \leq h(a) + h(b)$ and if b divides a, then $h(a/b) \leq h(a) + 2h(b)$.*

Proof. The multiplicative result is straightforward. Let $h_1 = h(a)$, $h_2 = h(b)$, and let the degrees of a and b be m and n respectively. Then the weight of the coefficient of s^{m+n-k} in the product is bounded by

$$\sum_{i=0,\ldots,k} h_1^i h_2^{k-i} \quad < (h_1 + h_2)^k \tag{3.29}$$

and so the product has height bounded by $(h_1 + h_2)$.

The quotient of a and b is a polynomial of degree $q = m - n$. To bound the height of the quotient, we use the matrix form for polynomial division (3.12), and we seek a bound on the weight of the subdeterminants of this matrix. The coefficients of (3.12) are bounded by:

$$\begin{pmatrix} 1 & h_1 & h_1^2 & \ldots & h_1^{q-1} & h_1^q & \ldots \\ 1 & h_2 & h_2^2 & \ldots & h_2^{q-1} & h_2^q & \ldots \\ 0 & 1 & h_2 & \ldots & h_2^{q-2} & h_2^{q-1} & \ldots \\ \vdots & \vdots & \vdots & & \vdots & \vdots & \\ 0 & 0 & 0 & \ldots & 1 & h_2 & \ldots \end{pmatrix} \tag{3.30}$$

We make several observations about the upper-left $k \times k$ submatrices of Q. Firstly, the determinant of such a matrix has total degree $k - 1$ in h_1 and h_2. This is easily proved by noticing that the leading diagonal has degree $k - 1$, and that degree is unaffected by interchanging the column indices of a pair of entries. Secondly, this determinant has at most 2^{k-1} terms (i.e. distinct powers of h_1, h_2). To see this, imagine computing the determinant by computing all the terms in all the $k \times k$ determinants of the first k columns. Suppose there are 2^{k-1}, and choose a term Q_l. Then for the $j+1^{st}$ column, there are $j+2$ non-zero entries (from the first $j+2$ rows), but j of these have already been used in the term Q_l, so we can choose exactly two non-zero elements of this column to give us two new terms. Thus there are at most 2^k terms, and the result follows by induction.

Now the coefficient of $h_1^i h_2^{k-i-1}$ in the $k \times k$ subdeterminant is found from the minor of the $i+1^{st}$ element of the first row. This minor can be split into an $i \times i$ unit diagonal matrix in the upper left, and a $(k-i-1) \times (k-i-1)$

matrix of the same form as Q but containing only powers of h_2 in the lower right. The weight of the determinant of this minor is therefore bounded by $h_2^{(k-i-1)}2^{(k-i-1)}$. So we have the following bound for the weight of the $k \times k$ subdeterminant:

$$\sum_{i=0,\ldots,k} h_1^i h_2^{(k-i-1)} 2^{(k-i-1)} < (h_1 + 2h_2)^k \tag{3.31}$$

and since this subdeterminant is the coefficient of s^{q-k} in the quotient, it follows that the height of the quotient is $(h_1 + 2h_2)$. □

Now if $p(x)$ is the polynomial:

$$p(x) = p_0 + p_1 x + \cdots + p_n x^n \tag{3.32}$$

then the following bounds hold:

Lemma 3.3.2 *If ξ is any (complex) root of $p(x) = 0$, then $|\xi|$ is bounded above by:*

$$|\xi| < 1 + \frac{\max(|p_0|, \ldots, |p_{n-1}|)}{|p_n|} \tag{3.33}$$

and if $\xi \neq 0$, then the following lower bound holds:

$$|\xi| > \frac{|p_0|}{|p_0| + \max(|p_1|, \ldots, |p_n|)} \tag{3.34}$$

Corollary 3.3.3 *If $p(x)$ is an integral polynomial of weight $w(p)$, ξ a non-zero root of $p(x) = 0$, then ξ satisfies:*

$$\frac{1}{w(p)} < |\xi| < w(p) \tag{3.35}$$

These and other bounds may be found in [Mig]. Now for the gap theorem, we consider the class $\wp(d, c)$ of polynomials of degree d and maximum coefficient magnitude c.

Theorem 3.3.4 (Gap Theorem) *Let $\wp(d,c)$ be the class of polynomials of degree d and coefficient magnitude c. Let $f_1(x_1,\ldots,x_n),\ldots, f_n(x_1,\ldots,x_n) \in \wp(d,c)$ be a collection of n polynomials in n variables which has only finitely-many solutions when homogenized. If $(\alpha_1,\ldots,\alpha_n)$ is a solution of the system, then for any j either*

$$\alpha_j = 0 \quad or \quad |\alpha_j| > (3dc)^{-nd^n} \tag{3.36}$$

Proof We homogenize the system by adding the homogenizing variable x_0. We then add the linear form $\sum v_i x_i$ and compute the u-resultant, which is a polynomial in $v_0,\ldots,v_n$. We assume for simplicity that $v_i = 0$ for $i \neq 0, j$, so that the resultant is an integral polynomial in v_0, v_j. Setting $v_0 = 1$, we obtain a polynomial whose roots are the values of α_j in the solution set. If we can bound the weight of the resultant, using the corollary to the last lemma, we obtain an immediate lower bound on $|\alpha_j|$. From method 2 equation (3.11), we see that the resultant is the product of $\bar{a}^k_k$ for $k = 1,\ldots,n+1$ divided by the product of the $\bar{a}^k_{k-1}$ for $k = 1,\ldots,n+1$. Now the matrices A^k all have size:

$$N = \binom{nd+1}{n} < \frac{(dn+1)^n}{n!} < (de)^n \tag{3.37}$$

where e is the base of natural logarithms. This follows from Stirling's approximation for factorial. Now $\bar{a}^k_k$ is the characteristic polynomial of matrix of size at most $Nk/(n+1)$, while $\bar{a}^k_{k-1}$ comes from a matrix of size $N(k-1)/(n+1)$. From the expression for height of a characteristic polynomial we have $h(\bar{a}^k_k) = Nck/(n+1)$ and $h(\bar{a}^k_{k-1}) = Nc(k-1)/(n+1)$. From the linear growth of height with multiplication and division we obtain:

$$h(\bar{R}) < \sum_{k=1,\ldots,n+1} \frac{3Nck}{(n+1)} \approx \frac{3}{2} Ncn \tag{3.38}$$

and since $\bar{R}(u_1)$ has degree d^n in u_1, the weight of the resultant (the constant term of $\bar{R}$) is

$$(3/2Ncn)^{d^n} < (3dc)^{nd^n} \tag{3.39}$$

and so the non-zero roots α_j of the resultant satisfy $|\alpha_j| > (3dc)^{-nd^n}$. □

This result implies that we can represent algebraic numbers defined by such a system using $O(nd^n \log(dc))$ bits, since the weight of the resultant also bounds the maximum size of a solution. In fact, the bound on the base of exponential growth, d, is tight because there exist systems having gaps of $d^{n-1} \log c$ bits. e.g. Consider $Lx_1^2 - x_1 = 0$, L a large integer, and the equations $x_i = x_{i-1}^d$ for $i = 2, \ldots, n$. Then $x_n = (1/L)^{d^{(n-1)}}$ is a solution.

This gap bound is the first single-exponential bound on the number of bits needed to represent algebraic numbers defined by systems of polynomials. It improves and generalizes the bound of Hong [Hon], who gave a gap theorem for the special case of triangular systems of polynomials.

3.4 Solution of Polynomial Systems

Algorithm 3.1
Input: *non-homogeneous polynomials $f_1, \ldots f_r$ in variables $x_1, \ldots x_{r-1}$, or homogeneous $f_1, \ldots f_r$ in variables $x_1, \ldots x_r$.*
Output: *a necessary and sufficient condition (in the form of an integral polynomial in the coefficients of the f_i) that the homogeneous (or homogenized) system have a solution over the complex numbers.*

Description Let d_i be the degree of f_i. We use method 2 above to compute the resultant of the system. This requires the computation of the characteristic polynomials of r matrices of size at most

$$N = \binom{r + d - 1}{r - 1} \tag{3.40}$$

where $d = 1 + \sum(d_i - 1)$ as before. Each characteristic polynomial can be computed using $O(N^4)$ operations on elements of the coefficient ring, and the maximum degree of any intermediate result in the polynomial coefficients is N. The $2r$ polynomial divisions require at most $O(N^2)$ operations each on the coefficient ring, and so the overall number of operations on the coefficient ring is $O(rN^4)$ (naively). Once again the maximum degree of intermediate values in the coefficient ring is N. Notice that we cannot compute the time complexity of the method without knowing the cost of multiplication and addition on elements of the coefficient ring as a function of their degree, e.g.

this function is pseudo-linear for integers, at least quadratic for polynomials in two variables etc.

On the other hand we can use method 1, which involves the computation of determinants of submatrices of the A^k, and therefore has a complexity of $O(rN^3)$ operations.

For the simple case where all the d_i are equal, the above expressions may be simplified, since $N = \binom{rd_1}{r-1} = O((d_1e)^r)$ from Stirling's approximation. Thus the number of coefficient operations is $O(r(d_1e)^{4r})$ for method 2, and $O(r(d_1e)^{3r})$ for method 1.

Algorithm 3.2
Input: *Homogeneous polynomials* $f_1, \ldots f_{r-1}$ *of degree* d *in* r *variables* $x_1, \ldots x_r$ *which have a finite number of distinct solutions. The solution rays are assumed to have distinct* x_1/x_r *ratios (this occurs with probability one after a change of coordinates).*
Output: *A list of the real root vectors* $(s_1, \ldots s_r)$, $s_r \neq 0$, *which are solutions of the system.*

Description We use the u-resultant defined in the last section. However, rather than computing the complete u-resultant as a polynomial in $v_1, \ldots, v_r$, we compute several resultants in three indeterminates, and then match solutions by x_1/x_r ratios. First we form $r-2$ resultants R_i, where each R_i is the resultant of the original polynomials and a linear polynomial

$$l_i = v_1x_1 + v_rx_r + v_ix_i \qquad \text{where } 2 \leq i \leq r-1 \tag{3.41}$$

Each R_i is a homogenous polynomial in three variables v_1, v_r, v_i, and it factors into linear factors $(\alpha_k v_1 + \beta_k v_i + \gamma_k v_r)$ where $(\alpha_k, \beta_k, \gamma_k)$ are the (x_1, x_i, x_r) coordinates of a solution ray. That is, the linear factors of R_i are the specializations of the linear factors of R, and they give only the (x_1, x_i, x_r) coordinates.

Since R_i is homogeneous, We can set $v_r = 1$ without affecting the factorization of R_i, giving a non-homogeneous polynomial in two variables. We could then apply a multivariate factorization algorithm, such as Kaltofen's, [Kal] to find all the factors of R_i. Instead, we use a simpler factorization scheme, described next.

We divide $R_i(v_1, v_i, v_r)$, as a polynomial in v_r, by the linear polynomial $(\alpha v_1 + \beta v_i + v_r)$, where the coefficients α and β are *indeterminates*. The α values of the factors that divide R_i now give the x_1/x_r ratio of the solution rays. The division produces a symbolic remainder $Q(v_1, v_i)$, a homogeneous polynomial in v_1 and v_i, whose coefficients are polynomials in α and β. Now the remainder $Q(v_1, v_i)$ must vanish if the division is exact, that is, all the coefficients of $Q(v_1, v_i)$ must vanish. Each of these coefficients is a polynomial of degree at most N in α and β.

It is not difficult to show that the coefficient c of v_1^N in $Q(v_1, v_i)$ is a polynomial in α only, and is not identically zero as long as there is some term in R_i which is not a multiple of v_i. To see this, we notice that c is the specialization $Q(1,0)$, i.e. $v_1 = 1$, $v_i = 0$. Another way to compute $Q(1,0)$ is to set $v_i = 0$ and divide $R_i(v_1, 0, v_r)$ by the linear factor $(\alpha v_1 + v_r)$ as polynomials in v_r. Then the remainder vanishes only if this linear factor divides $R_i(v_1, 0, v_r)$ for all α, and this can only happen if $R_i(v_1, 0, v_r)$ is indentically zero. In other words, every term in R_i is a multiple of v_i. If this is the case, then R_i has v_i as a factor, and we add it to the list of factors and divide by it until R_i is no longer a multiple of it, call the result R_i^{red}. Eventually, the coefficient of highest degree in v_1 of R_i^{red} is a polynomial in α only (of degree at most N), which must be zero for values of α such that α the coefficient of v_1 in a linear factor.

By finding approximate roots of this polynomial, we can find all possible values of α that can occur in factors of R_i of the form $(\alpha v_1 + \beta v_i + v_r)$. Similarly, from the coefficient of v_i^N in Q, we can find all the values of β. To find factors of R_i, we enumerate all N^2 pairs of approximate roots α and β and substitute each pair into the other coefficients of Q. These other coefficients will vanish if and only if the pair occur together in a factor. We test for the vanishing of the coefficients using the gap theorem. In this way, we can compute all of the approximate factors of $R_i(v_1, v_i, v_r)$ of the form $(\alpha v_1 + \beta v_i + v_r)$.

To find factors which are not of the form $(\alpha v_1 + \beta v_i + v_r)$, i.e. those having the coefficient of $v_r = 0$, we do the symbolic division of R_i by the polynomial $(v_1 + \beta v_i + \gamma v_r)$, but now treating them as polynomials in v_1. Then we repeat the steps outlined in the previous paragraphs for the remainder polynomial $Q(v_i, v_r)$. This picks up the factors having coefficient of v_1 non-zero. After performing all these steps for factors with coefficient of v_r non-zero, or coefficient of v_1 non-zero, the only remaining factors are those where both coefficients are zero. But these factors are just multiples of v_i, and will be detected as described in the last paragraph (the resultant will be a multiple of v_i).

Having found approximate factorizations of the R_i, we identify them by α values. Since the value of α_k in some factor $(\alpha_k v_1 + \beta_k v_i + v_r)$ gives the ratio of x_1/x_r of the kth solution ray, if some factor $(\alpha_l v_1 + \delta_l v_j + v_r)$ of R_j has $\alpha_k = \alpha_l$, then this second factor comes from the same solution ray. We conclude that the coordinates of this ray are $x_1 = \alpha_k = \alpha_l$, $x_i = \beta_k$, $x_j = \delta_l$, $x_r = 1$. Doing this for each R_i, we get all of the coordinates of all solution rays for which the coordinate of x_r is non-zero. There still remain the solutions for which x_r is zero, but by assumption the ratios x_1/x_r are unique, and there will be only one solution having x_r coordinate zero.

We next analyse the complexity of the method. If the polynomials have degree d, then $N = d^r$ is the degree of the resultant in the v_i. The coefficients of Q also have degree at most N in α and β. Computation of R_i for each i using method 1 can be done with $O(r(de)^{3r})$ operations on homogeneous polynomials in v_1, v_i, v_r. Each operation can be done in $O(rd^{2r} \log d)$ time using FFT multiply, so the resultant computation overall takes time $O(r^2(de)^{5r} \log d)$. The division of each R_i by the symbolic linear factor takes time $O(d^{4r})$ naively.

The test for vanishing of some coefficient of Q involves three polynomials, the coefficient itself, and the polynomials which define α and β. The length of the (integer) coefficients of Q, as a polynomial in v_1, v_i, α and β is at most a constant multiple of the coefficient length of R_i. From this it follows that the gap size for the value of the test is $O(rd^{3r} \log dw)$ bits, where w is the maximum absolute value of any coefficient of any f_i. By simple error analysis, using the upper bounds for α and β, and the fact that coefficients of Q have degree d^r, we need $O(rd^{3r} \log dw)$ bits of accuracy in the value of each root to ensure that the error in the value of R_i is within the gap.

Once we have a sufficiently good approximation to a root, Newton's method can be used to isolate the root to the above accuracy in time $O(r^3 d^{3r} \log^3 dw)$. Evaluation of a coefficient of Q for each solution β_k takes time $O(r^2 d^{3r} \log^2 dw)$. (Both of the previous two bounds make use of a simple divide-and-conquer scheme for polynomial evaluation, and rounding to $O(rd^{3r} \log dw)$ bits). To find all three-element solution rays (solution rays for some R_i) takes time $O(r^2 d^{5r} \log^2 dw)$. Once we have these for each R_i, we can immediately obtain all n-element real root vectors by comparing their x_1-coordinates. The cost for evaluation of coefficients of Q dominates the time for computation of the resultants R_i and the remainders Q, and the overall cost for this algorithm is $O(r^3 d^{5r} \log^2 dw)$ binary operations.

Chapter 4

The Roadmap Algorithm

In this chapter we describe the roadmap algorithm, which solves the generalized movers' problem in single exponential time, with exponent equal to the number of degrees of freedom of the robot. The algorithm assumes that the set of free configurations of the robot is defined as a semi-algebraic set, and that this set is closed and compact. We also need the polynomials which define the semi-algebraic set to be in general position (we will be more precise about what we mean by this later). For polynomials that are not in general position, we show that almost any perturbation of the constant coefficients of the polynomials achieves this. The semi-algebraic set defining free-space could be generated from a description of the robot and its environment by the methods of chapter 2, or by any other derivation that leads to a semi-algebraic description of free-space.

Given the set S of free configurations, the algorithm produces a one-dimensional semi-algebraic subset $R(S)$ of S called the roadmap of S. The roadmap consists of algebraic curve segments which meet at points, and the algorithm produces an adjacency graph for $R(S)$ whose vertices represent points of $R(S)$ and whose edges represent the curve segments. The roadmap has the important property that it is connected within each connected component of S. We also show that an arbitrary point in S can be linked to the roadmap in linear time. Thus if any path exists between two points in S, then we can find a path by linking the two points in to the roadmap. If such a path exists, then the two points must lie in the same connected component of S. The linking algorithm will link them to the same component of the roadmap, and we can check whether they are in the same component by searching the adjacency graph of the roadmap.

In section 4.1 we give a non-trivial example of the operation of the algorithm. The example is easy to visualize, and doesn't require any knowledge of topology or geometry to understand. While the example is three-dimensional, it should give the reader some idea of how the algorithm works in higher dimensions.

In section 4.2, we define some basic concepts and results from differential topology which will be used throughout the chapter. Although all the concepts used in this chapter are defined here, some familiarity with differential topology at the level of critical points of maps, transversality and Sard's theorem is extremely helpful. This material should be available in any textbook on the subject, and we recommend in particular the books by Guilleman and Pollack [GP] and Golubitsky and Guilleman [GG], whose notation we loosely follow.

In section 4.3 we introduce stratified sets, which we feel are a very natural abstraction for representing the kinds of geometric objects which are encountered in robotics. Stratified sets may be thought of as generalized or curved polyhedra. They are applicable not only to "well-behaved" algebraic and semi-algebraic sets, but also to singular sets. Currently, they are not covered in basic textbooks, although a good reference is the monograph by Gibson, Wirthmüller, Du Plessis and Looijenga [GWDL]. The first chapter of that work describes the basic properties of stratified sets, while the second deals with controlled vector fields which allow global geometric analysis to be reduced to local analysis of critical points.

Section 4.4 describes silhouette curves from which the roadmap is constructed. This section takes further concepts from differential topology and makes them concrete by expressing them in terms of embedded manifolds and jacobians. The objective is to give constructive or algorithmic descriptions, in preparation for the description of the roadmap algorithm itself in section 4.5.

Section 4.5 defines the basic roadmap, linking curves, and the full roadmap. In this section we define an extension of the stratification of S into sign invariant sets which includes a stratification of the silhouette as a subset. The basic roadmap has a simple recursive definition based on this stratification. The recursive call at a critical value of a real-valued map is reminiscent of cell adjunction at critical values in Morse theory. Linking curves are defined using the basic roadmap, and allow a more efficient kind of roadmap, called the full roadmap. Finally the algorithm for full roadmaps is described in detail in section 4.6.

4.1 An Example

The roadmap algorithm finds the connected components of a high-dimensional set S by computing a one-dimensional skeleton or roadmap $R(S)$. The skeleton has the property that it is connected within each connected component of S. This one-dimensional set, even though it has algebraic curves joining its vertices, is much easier to work with than the original set S. We can explicitly compute the connectivity of $R(S)$, and it can be naturally represented as a graph whose vertices are the vertices of $R(S)$, and whose edges are the algebraic curve segments joining vertices. To find a path between two arbitrary points in S, all we need to do is to find a path from each point to some nearby point on the skeleton. Then if the two points lie in the same connected component of S, there will exist a path between them through the skeleton, which can readily be checked from the graph representation of $R(S)$.

The rationale behind the algorithm for computation of $R(S)$ follows: The main idea of the algorithm is to reduce the dimension of the problem by one by taking *slices* through the original set S with a movable plane P_c (P_c is a hyperplane in higher dimensions) as shown in figure 4.1. In the figure, S is an ellipsoidal solid with a cylindrical hole drilled in it, not quite in a vertical direction. The plane P_c in the example is defined by the equation $x = c$. While S is three-dimensional, every slice $S \cap P_c$ is two-dimensional. We get a useful reduction in dimension if we can somehow reduce the problem of finding a path in S to several subproblems of finding paths in the slices $S \cap P_c$.

In order to do this, we should define the skeleton in such a way that we can always find a path from an arbitrary point in $S \cap P_c$ to some point in $R(S)$ *without leaving* $S \cap P_c$. This gives us a requirement that every connected

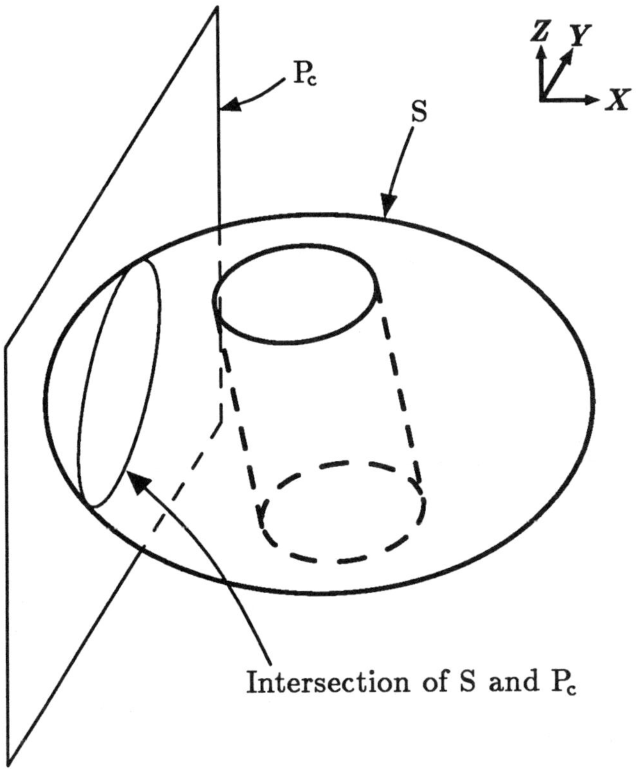

Figure 4.1: The set S, and a slice through it by the plane P_c

component of $S \cap P_c$ should contain a point of $R(S)$. Here we choose, arbitrarily, some direction in the plane P_c, say the y-direction, and require that $R(S)$ contain all the *locally extremal* points, in the y-direction, of $S \cap P_c$. A locally extremal point in the y-direction is a local maximum, minimum or inflection point in the y-direction. See figure 4.2. There is guaranteed to be such an extremal point in each connected component of $S \cap P_c$, since y attains maximum and minimum values somewhere in each component, and the points where it does so will be extremals. Strictly speaking, this is only true if the set S is compact, a requirement we mentioned in the introduction to this chapter. If we choose our coordinates carefully, (in fact almost any

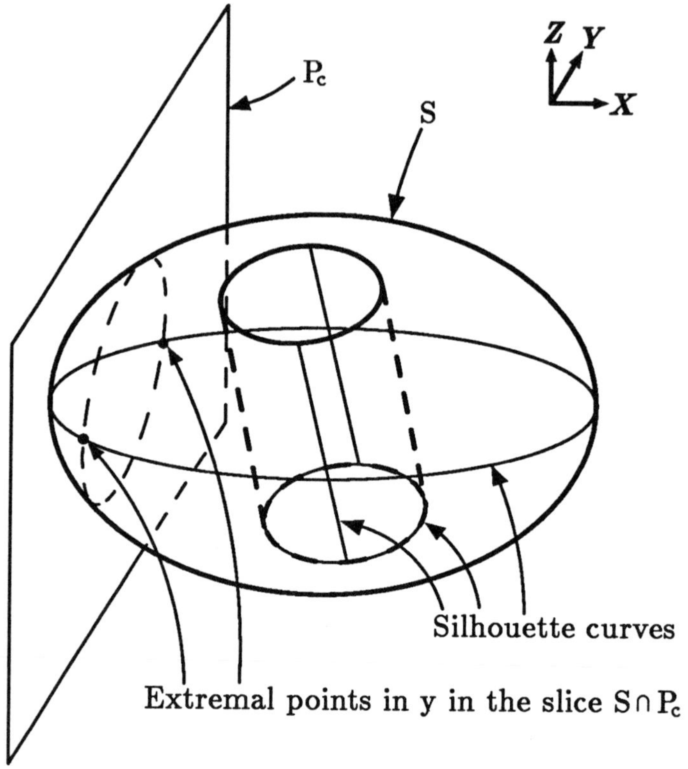

Figure 4.2: Locus of extremal points of slices $x = c$ as c is varied.

choice will do) we will always have just a finite number of extremal points in every slice.

Now as we vary c, the extremal points trace out curves, as shown in figure 4.2. These curves form the backbone of the skeleton $R(S)$, and are called *silhouette* curves. A second example appears in figure 4.10, which shows the silhouette of a torus. The silhouette curves are so called because they are precisely the curves that project down to give the silhouette of the set S, when S is projected onto the x-y plane. Even in higher dimensions, the silhouette curves are the curves that project down to give the x-y silhouette of S, where x is the direction normal to the hyperplane P_c, and y is the direction in which extremals are computed. If the coordinate directions are suitably chosen, the silhouette curves are smooth and one-dimensional within each smooth piece of S, independent of the dimension of S (except for points, whose silhouette is the points themselves, and is therefore zero-dimensional).

So silhouette construction can be visualized by imagining the set S to be made of translucent material, such that when S is viewed from a point at infinity, the contours that the viewer sees on the surfaces of S are the silhouette curves. For a sphere, one sees a great circle of points on its equator, while for a general view of a cylinder, one sees two parallel lines that share a plane with the axis of the cylinder.

But the silhouette curves are only part of the answer. Even in our simple example of figure 4.2, the silhouette is not connected, even though the set S is. One connected component of the curves passes around the perimeter of the ellipsoid. The second component includes the circular ends of the cylindrical hole, and the two lines on the boundary of the cylindrical hole, parallel to its axis. To remedy the lack of connectivity, we go back to our original device of slices $S \cap P_c$. As we increase c from some value c_0, the extremal points in the slice trace out smooth curves, so it would seem that if the silhouette were connected for $c < c_0$ it should remain connected as we increase c. But notice from figure 4.2 that new extremals appear (or disappear) in certain slices. The values of c at which this happens are called critical values. The new extremal points themselves are called critical points. A point on a silhouette curve is a critical point if the tangent to the curve at the point lies in the plane P_c.

So while the connectivity of the slice $S \cap P_c$ may change when c passes through some critical value, it remains the same as c varies over any interval that does not contain a critical value (in particular, in the open interval between two consecutive critical values). When the slice $S \cap P_c$ *does* pass through a critical point p, we can try to connect p to the rest of the silhouette with a path that lies *within the slice* $S \cap P_c$. Remember that the silhouette was defined so that this should always be possible. The process of joining a critical point to the existing silhouette is called linking. The simplest way to link a new point is to call the roadmap algorithm *recursively* on the slice $S \cap P_c$.

So our definition of the algorithm is recursive, and the proof that it succeeds is by induction on dimension. We assume that the algorithm correctly computes the roadmaps of slices through all the critical points, and that these roadmaps include the critical points. This implies that as c is increased, whenever a new extremal point appears in some critical slice, it will be connected (by a path within that slice) to the rest of the silhouette. Since connectivity is unchanged as c is increased to the next critical value, the roadmap, now including the new extremal curves, remains connected. If the dimension of S is r, and if our claim that the algorithm works on the $r-1$ dimesional slices is correct, we have shown that it also works in dimension r. It is easy to compute a roadmap if S is one-dimensional, since we can just return the set S. So by induction on dimension, the algorithm works for any r.

There is really a second induction step buried in the above argument. We use the notation $S|^{\leq c}$ to denote the set $S \cap (x \leq c)$. What we are claiming is that $R(S)|^{\leq c}$ is connected within each connected component of $S|^{\leq c}$. The inductive hypothesis is that this claim is true for some value $c = c_0$. Then it remains true as c is increased, until c reaches the next critical value c_1, with associated critical point p. If the roadmap algorithm works on the slice $S \cap P_{c_1}$, then even if a new component of silhouette appears, or if several components of S come together at p, all components of the silhouette will be joined by the recursively computed roadmap. Thus $R(S)|^{\leq c_1}$ is connected within each connected component of $S|^{\leq c_1}$. The base case is simply where c is small enough that $S|^{\leq c}$ is empty, in which case the inductive hypothesis is vacuously true. It will be shown later that for semi-algebraic sets S, the number of critical values is finite. So, by induction on the critical values, $R(S)$ is connected within each connected component of S.

Figure 4.3 shows the set S with silhouette curves and the roadmap of a slice through *one* of the critical points. The addition of this roadmap is enough to connect the two components of the silhouette. There are three other critical points, and three similar recursive roadmaps that constitute the full roadmap of S. Each slice $S \cap P_c$ at a critical value of c consists of a two-dimensional interior region bounded by one-dimensional curves. The roadmap of each *slice* consists of the silhouette of the slice together with recursively computed slices (now one-dimensional) through critical points of

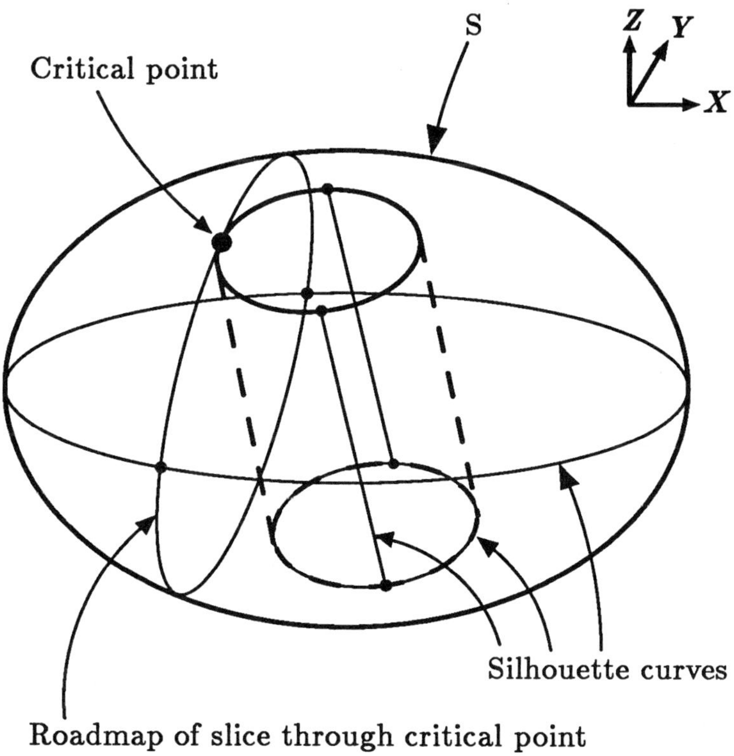

Figure 4.3: The silhouette of S and the recursively computed roadmap of the slice through one of the critical points.

the silhouette. As was the case for the set S, the silhouette of each slice is found by sweeping a plane (in this case a line) through the slice, and following the extremal points in some direction in the plane (since the plane here is a line, there is no choice of this direction). The boundary curves of the slice are automatically part of the silhouette of the slice, and in the example, no further recursive slices need to be computed. So the *roadmaps* of the slices in the example consist only of these elliptic bounding curves. An example of a complete roadmap is given in figure 4.11, the roadmap of the torus.

What we have described above is called the *basic roadmap* of S. The objective of recursive computation of roadmaps of slices was simply to link new critical points to the existing roadmap, but it is not necessary to compute the complete roadmap of the slice to do this. The *full roadmap* is a refinement of the basic algorithm where the linking curves are subsets of the roadmaps of slices, and are much less expensive to compute. In fact the linking curves described later in this chapter can be computed in time linear in the number of polynomials defining S, even in higher dimensions.

The rest of the chapter develops the mathematical tools we need to actually compute the roadmap. In section 4.3, stratifications are described. In our description above, we had to compute the extremal points in y of each slice. To do this it is necessary that the set S be decomposed into smooth pieces, i.e. surfaces or "manifolds" of some dimension. Then an extremum in y can be defined as a point where all of the tangents to the surface have y-component zero. The decomposition into smooth pieces is called a *stratification* of S. In section 4.3 we show how guarantee that all of the strata in the stratification are smooth manifolds, by making arbitrarily small perturbations in the constant coefficients of the polynomials defining S. This is the first "general position lemma", and it makes use of transversality as a simple condition which guarantees a regular stratification.

Section 4.4 deals with computation of the silhouette. A large part of this section is devoted to finding the conditions on the coordinate directions (the x and y directions in the above example) so that the silhouette curves are one-dimensional and non-singular. The second general position lemma, proved in this section, shows that all most any linear map into $\Re^2$ will work, so that almost any choice of a pair of directions instead of x and y will work for the example above. The proof of the result makes use of jet transversality, which makes it perhaps the most difficult section of the chapter. However, the result itself is quite intuitive, and the reader does not need to understand this section to follow the rest of the chapter. Also proved in this section is a result on tubes around varieties. This last result is not central to the construction of the roadmap, it simply shows that critical points of arbitrary varieties can be computed from the critical points of hypersurfaces. This gives a simple and efficient method for computation of critical points, but it is not the only method. The result is only used in the description of the roadmap algorithm in section 4.6.

Section 4.5 describes the roadmap in detail. It gives precise definitions of the basic and full roadmaps, and proofs that they have the right connectivity. These proofs use roughly the same argument given above to show that $R(S)$ is connected within each connected component of S. The linking curves, used to link new critical points in the full roadmap, are described in detail in this chapter, and are shown to be of three types. Finally, section 4.6 develops the algorithm itself. The algorithm is defined bottom-up, starting with algorithms for computing critical points, then for computing the projections of curves, and the adjacencies of curves and points. The basic roadmap is built up from these algorithms, and the linking curve algorithm is very similar to it. Finally, the full roadmap algorithm is described and a detailed complexity bound is given for it.

4.2 Preliminaries

The input to the roadmap algorithm will be a semi-algebraic set S defined by polynomials with rational coefficients. The r-dimensional configuration space is represented by r real values, so the polynomials defining the set come from the ring Q_r of polynomials in r variables with rational coefficients, $Q_r = Q[x_1, \ldots, x_r]$. We now make the notions of algebraic and semi-algebraic set formal.

Definition. An *algebraic set* in $\Re^r$ is the set of simultaneous zeros of a system of polynomials in Q_r.

Definition. A *semi-algebraic set* $S \subseteq \Re^r$ defined by the polynomials $F_1, \ldots F_n \in Q_r$ is a set derived from the sets

$$S_i = \{x \in \Re^r \mid F_i(x) > 0\} \tag{4.1}$$

by finite union, intersection and complement.

As was mentioned in the introduction, semi-algebraic sets have more descriptive power than algebraic sets. In particular, it is possible to define the interior or exterior of an object as a semi-algebraic set, but not as an algebraic set.

Example 4.1 For example, the algebraic set defined by $x^2 + y^2 = 1$ in $\Re^3$ is a cylinder, while to define the interior of this cylinder, we need an inequality $x^2 + y^2 < 1$. To define a finite closed cylindrical solid we use two planar constraints in addition to the cylindrical constraint, giving us the semi-algebraic set defined as:

$$(x^2 + y^2 \leq 1) \wedge (z \leq 1) \wedge (z \geq -1) \tag{4.2}$$

While the predicates which define semi-algebraic sets are a compact and convenient representation for the sets, they lack any geometric information. In fact, it may not even be clear from the predicate if the set is empty or not. In order to compute geometric properties of the sets, we will be making use of a partition of the sets into geometrically simple pieces called *manifolds.*

Definition. A subset M of $\Re^r$ is a differentiable *manifold* of dimension m (an m-manifold) if every point in M has a neighborhood which is diffeomorphic to $\Re^m$. Thus an open subset of a manifold is also a manifold. By convention, the empty set has dimension -1.

Intuitively, an m-manifold corresponds to the notion of a "smooth m-dimensional surface". Notice that objects with corners or edges fail to be manifolds, so that a closed cube, or the closed cylindrical solid of example 4.1 fail to be manifolds. This is because boundary points of these objects have neighborhoods which are diffeomorphic only to a closed subset of some $\Re^m$. The surface $x^2 + y^2 = 1$ in $\Re^3$ is a manifold however.

The local diffeomorphisms that exist at each point in a manifold allow us to define local coordinates near each point, which are the image under the diffeomorphism of the usual coordinates on the space $\Re^m$. Let $x \in M$ be a point with neighborhood $U \subset M$, and let $\phi_x : U \to V$ be a diffeomorphism, where $V \subset \Re^m$ is open. The maps ϕ_x are called coordinate charts. We will assume that the charts are *adapted* to x so that $\phi_x(x) = 0$. Since the range of ϕ_x is $V \subset \Re^m$, it may be thought of as m "coordinate functions". Rather than explicitly using charts, we will sometimes speak of "local coordinates" about a point $x \in M$, which actually means the values of the m coordinate functions.

Definition For a map $f : \Re^m \to \Re^n$ the *differential* of f relates infinitesimal changes in the domain of f to the corresponding changes in the value of f. The differential of f at x is denoted df_x, and is given by the jacobian:

$$df_x = \begin{pmatrix} \frac{\partial f_1}{\partial x_1}(x) & \dots & \frac{\partial f_1}{\partial x_m}(x) \\ \vdots & & \vdots \\ \frac{\partial f_n}{\partial x_1}(x) & \dots & \frac{\partial f_n}{\partial x_m}(x) \end{pmatrix} \tag{4.3}$$

Definition We can now define the space of vectors tangent to a manifold using the differential of an inverse chart. Let ψ be the inverse of the coordinate chart ϕ_x. We define the *tangent space* to M at x, denoted T_xM, as $d\psi_0(\Re^m)$.

The notion of differential can now be generalized to maps between manifolds. Let f be a smooth map of manifolds $f : M \to N$, where M and N have dimension m and n respectively. We define the differential of f at $x \in M$, (denoted df_x) as the induced map of tangent spaces $df_x : T_xM \to T_{f(x)}N$. Using local coordinates about x and $f(x)$, the differential is again given by the jacobian (4.3).

Once we have defined maps between manifolds, a natural question arises. Given a map $f : M \to N$, under what conditions is the preimage $f^{-1}(v)$ of some $v \in N$ also a manifold ? What about the preimage of some submanifold $V \subset N$? This brings us to the definition of critical points and transversality.

Definition. A point x in M is a *critical point* of $f : M \to N$ if the differential df_x is not surjective at x, i.e. if the jacobian (4.3) has rank less than $\dim(N)$ at x. We will call the set of all critical points of f its *critical* *set*, and denote it by $\Sigma(f)$. Points where df_x is surjective are called *regular points.*

The image of a critical point $x \in M$ is a critical *value* of f. Values $v \in \Re^n$ which are not critical values are called *regular* values. So a value v is regular if and only if all the points in $f^{-1}(v)$ are regular. In particular, v is a regular value if $f^{-1}(v)$ is empty.

Example 4.2 Let M be the unit sphere in $\Re^3$ defined by $x^2+y^2+z^2=1$, and let N be $\Re^2$. Let $f : M \to N$ be $\pi_{12}|_M$, that is, projection on x and y coordinates, with domain restricted to M, so that $f(x,y,z) = (x,y)$. For most points $p \in M$, the image of the tangent plane to the sphere at p covers the x-y-plane. At the equator however, the tangent plane contains a vertical vector, and projects down to a line in the x-y-plane, as shown in figure 4.4. Thus points on the equator are critical points, and they project to a circle of critical values. Notice that $\pi_{12} : \Re^3 \to \Re^2$, without its domain restricted, has no critical points.

The regularity of a map is precisely the condition we need to guarantee that the preimage of a point is also a manifold, in fact, we can even determine its dimension.

Theorem 4.2.1 (Preimage theorem.) *If v is a regular value of the smooth map $f : M \to N$ then the preimage $L = f^{-1}(v)$ is either empty or a sub-manifold of M of dimension* $\dim(M) - \dim(N)$. *Furthermore, we have a simple representation of T_xL,*

$$T_xL = \ker(df_x) \tag{4.4}$$

Since locally the tangent space to L is the set of differential displacements which do not change the value of f.

Next we ask when the preimage of a manifold $V \subset N$ is also a manifold. The condition we need is that the map f be transversal to V.

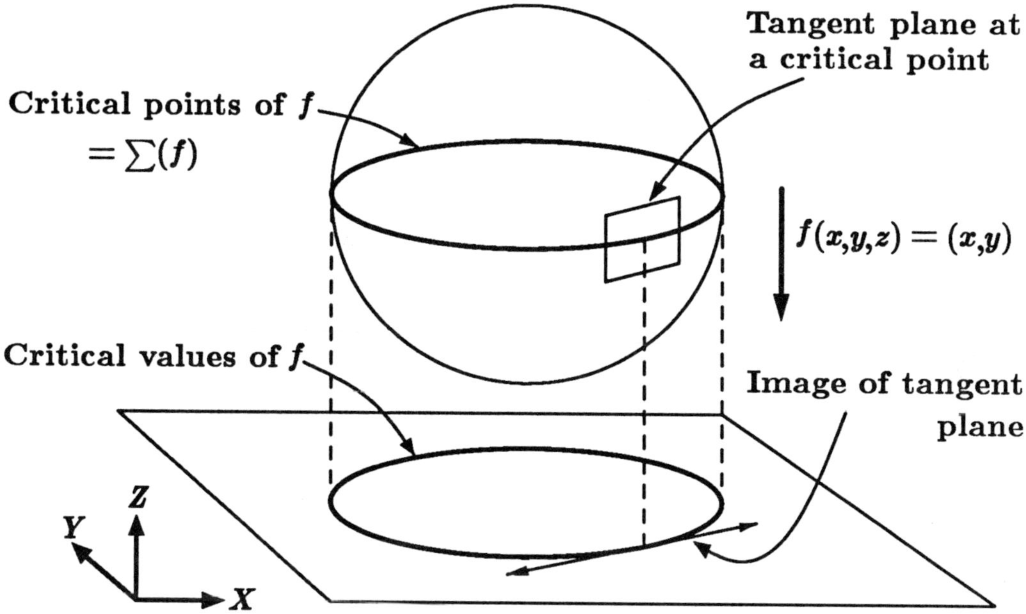

Figure 4.4: Critical points of the projection map from the sphere to $\Re^2$

Definition. Let $f : M \to N$ be a smooth map, and let V be a submanifold of N. We say that f is *transversal* to V if

$$df_x(T_xM) + T_{f(x)}V = T_{f(x)}N \tag{4.5}$$

at every point $x \in f^{-1}(V)$, and we denote this relationship by $f \pitchfork V$. The + here denotes sum as vector spaces, in other words, the sum of two spaces is the space spanned by all the vectors in them.

Transversality at a point is a weaker condition than regularity. If x is a regular point, then the image of T_xM under df_x must equal $T_{f(x)}N$, whereas for transversality we require only that the sum (as vector spaces) of $df_x(T_xM)$ and the tangent space $T_{f(x)}V$ equal $T_{f(x)}N$. So if every $v \in V$ is a regular value, then $f \pitchfork V$.

Theorem 4.2.2 (Generalized Preimage Theorem) *If $f \pitchfork V$, then the set $f^{-1}(V)$ is either empty or a manifold whose codimension is* codim(V). *(The codimension of a submanifold $V \subset N$ is* codim(V) = dim(N) − dim(V)*).*

Example 4.3 We take M to be the cylinder in $\Re^3$ defined by $x^2+z^2=1$, N to be $\Re^2$ and $f : M \to N$ to be the projection map π_{12} with domain restricted to M, so that $f(x,y,z) = (x,y)$. If $V \subset N$ is the line $y = 0$, we have the situation depicted in figure 4.5. Now $f \pitchfork V$ because *df* is surjective except at critical values, and at each point $y \in V$ which is critical value of f, the image of $T_{f^{-1}(y)}M$ under $df_{f^{-1}(y)}$ is orthogonal to T_yN, so these two vectors span the x-y-plane. The preimage $f^{-1}(V)$ is therefore a smooth manifold, in fact a circle in this case.

Example 4.4 Let M, N and f be as in the previous example. Suppose now we take V to be the circle in the x-y-plane defined by $x^2+y^2=1$, as shown in figure 4.6. Once again if y is a regular value of f, then all is well, but there are two critical values of f in V. At these two points, the image of the tangent space to M under *df* and the tangent line to V are both parallel lines, and they do not span the x-y-plane. Thus f fails to be transversal to V, and the preimage $f^{-1}(V)$ is a pair of ellipses which intersect at two points, so $f^{-1}(V)$ is not a manifold. It fails to be so precisely at those points x where $df_x(T_xM) + T_{f(x)}V \neq T_{f(x)}N$.

A map which is regular everywhere is given a special name.

Definition. A smooth map $f : M \to N$ is called a *submersion* if every $v \in N$ is a regular value of f.

For a map $f : M \to N$, let $U = f^{-1}(v)$ for some $v \in N$. We use the slice notation $f|_U : U \to N$ for the restriction of the domain of f to U. Now the following are equivalent:

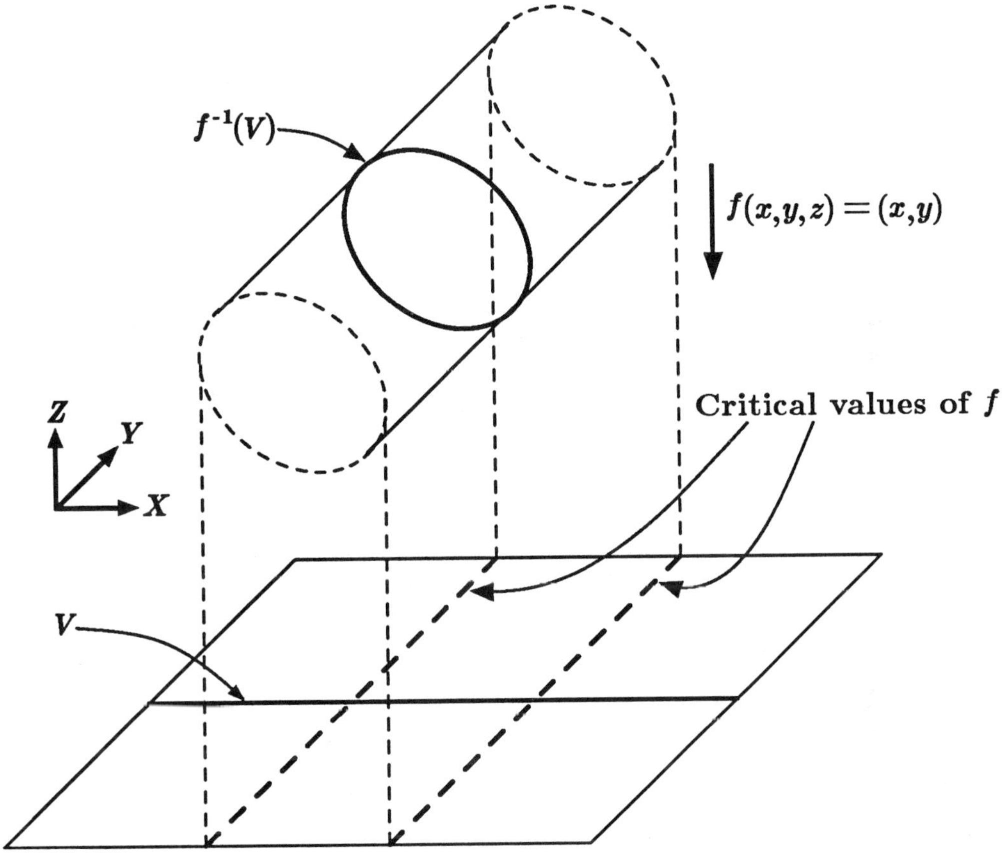

Figure 4.5: Example of a map f transversal to a submanifold V of the x-y-plane, the preimage of V is a circle in this case

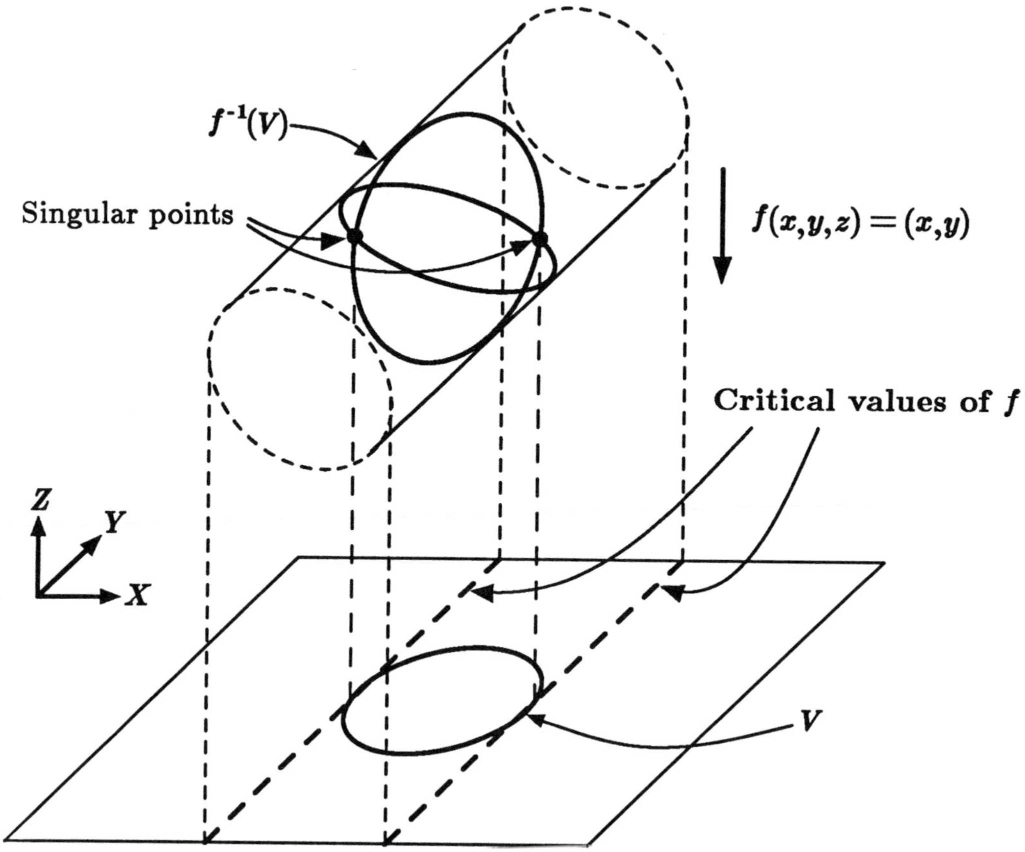

Figure 4.6: Example where f fails to be transversal to V. The tangent to V and the image of the tangent plane of M are parallel at critical values of f. $f^{-1}(V)$ is not a manifold in this case

(i) v is a regular value of f.
(ii) $f \pitchfork \{v\}$
(iii) $f|_U$ is a submersion.

and all of these imply that U is a manifold of codim $= \dim(N)$.

A major theorem from Analysis tells us that almost all values in the range of a smooth function are regular values. It will be used in two "general position" lemmas which we will use to avoid a variety of singular conditions.

Theorem 4.2.3 *(Sard): Let $f : M \rightarrow N$ be a smooth map of manifolds, and let C be the set of critical points of f in M, then $f(C)$ has measure zero in N. In particular, let $f : \Re^r \rightarrow \Re^n$ be a smooth map with $r \geq n$, then for almost all $v \in \Re^n$ the set $f^{-1}(v)$ is either empty or a manifold of dimension $r - n$.*

As we saw in the last 3 examples, the set of critical values of the projection maps f were all one-dimensional subsets of $\Re^2$ and thus had measure zero in $\Re^2$.

In the context of this chapter, where we will be dealing only with semi-algebraic sets and polynomial maps, the set of critical values will always be semi-algebraic. In this case, a measure zero subset will be a semi-algebraic subset of lower dimension. Intuitively, measure zero corresponds to the notion of probability zero, or zero volume.

The following result is readily derived from Sard's theorem, and will be used directly in the general position lemmas. A proof of both theorems may be found in [GG].

Lemma 4.2.4 *(Generic Map Lemma) Let M, N, B be smooth manifolds, and let V be a submanifold of N. Suppose $\Phi : M \times B \rightarrow N$ is a smooth map such that for every $b \in B$, $\Phi_b : M \rightarrow N$ is smooth, where $\Phi_b(x) = \Phi(x, b)$. If $\Phi \pitchfork V$, then $\Phi_b \pitchfork V$ for almost all $b \in B$ (all but a measure zero subset).*

In other words, if a parametrized class of mappings is transversal to V, then almost every map in the class is transversal to V.

4.3 Stratifications

Differential topology has traditionally concentrated on the study of global properties of differentiable manifolds, which look locally like euclidean spaces of fixed dimension. There are many important examples of sets which have such euclidean neighborhoods almost everywhere, but which fail to be manifolds. In particular, algebraic sets of dimension d may have singular subsets of lower dimension. Whitney [Wh57] in attempting to describe the geometry of algebraic sets, showed that a singular algebraic (or semi-algebraic) set of dimension d can be partitioned into a finite number of manifolds of dimensions ranging from 0 to d. This result has been extended to sub-analytic sets, and such a partition has come to be known as a stratification (or stratified set). Stratifications enjoy many of the useful properties of manifolds, and in particular they can be triangulated [Joh]. More recently, stratifications have played an important role in proofs of some fundamental stability theorems for maps between manifolds [GWDL].

Stratifications are highly applicable to computational geometry and robotics, since the class of sets which are easily described as stratified sets includes polyhedra, constructive solid geometric (CSG) models, and the transformed obstacles in the configuration space [Loz] of virtually any robot. The arrangement algorithm for hyperplanes of [EOS] can be viewed as a stratification procedure. So can the cell decomposition algorithm for semi-algebraic sets of [Col]. The latter is however extremely fine, and ill-suited to geometric applications, since it produces a double exponential (in the dimension of the space) number of strata. We will be here considering a coarsest possible stratification of semi-algebraic sets, which has a single exponential number of strata.

Definition A *stratification* $\underline{\underline{S}}$ of a set $S \subset \Re^r$ is a partition of S into a finite number of disjoint subsets S_i called *strata* such that each S_i is a manifold.

An additional condition is often used in the definition of a stratification. Let $\overline{V}$ denote the closure of V. A stratification $\underline{\underline{S}}$ satisfies the *Frontier Condition* if whenever $U \cap \overline{V} \neq \phi$ with U, V in $\underline{\underline{S}}$, then $U \subset \overline{V}$. In this case we say U is *adherent* to V. Adherence is a transitive relation, and defines a partial order on the strata of $\underline{\underline{S}}$. For this reason, it is sometimes denoted $U < V$. The theorems on stratifications require a certain joining property for adherent strata, which we give now:

The Whitney Conditions Let U and V be any two strata of a stratification $\underline{\underline{S}}$, such that $U \subset \overline{V}$. Let $x \in U$ and let (x_i) and (y_i) be any two sequences in U and V respectively which converge to x. The following are called the Whitney conditions:

(A) If $(T_{y_i}V) \rightarrow \tau$, then $T_xU \subset \tau$.
(B) If $(T_{y_i}V) \rightarrow \tau$ and $(\overline{x_iy_i}) \rightarrow l$, then $l \subset \tau$.

where $\overline{x_iy_i}$ is the line passing through x_i and y_i. Note that the second condition implies the first. A stratification which satisfies condition (B) is called a *Whitney stratification*, or sometimes a *regular stratification*. In this definition, we have not been precise about what it means for a sequence of tangent spaces to converge to a space τ, relying on the reader's geometric intuition. A precise definition involves a Grassmanian bundle of planes over the manifold, and can be found in [GWDL].

Example 4.5 The Cartan umbrella. Consider the algebraic set A in $\Re^3$ defined by $y^2 = zx^2$ and shown in figure 4.7. The set A can be stratified into two subsets, A_0, which is the z-axis, and A_1 which is $A - A_0$, as shown in figure 4.8(a). This stratification does not satisfy the frontier condition, nor does it satisfy Whitney's condition (A). To see this, consider a sequence of points in A_1 on the x-axis which converges to the origin. The limit of the tangent spaces of this sequence is the x-y-plane. But the tangent space to A_0 at the origin is the z-axis, which is not contained in the x-y-plane.

However, there is a finer stratification of A which does satisfy the Whitney conditions. We subdivide A_0 into three subsets where $z > 0$, $z = 0$ and $z < 0$, as shown in figure 4.8(b). Here the tangent space at the origin is empty, and is trivially contained in the limit of tangent spaces of points converging to x.

In [Wh57], Whitney showed that any algebraic set admits a stratification. In fact this paper preceded Whitney's development of regularity

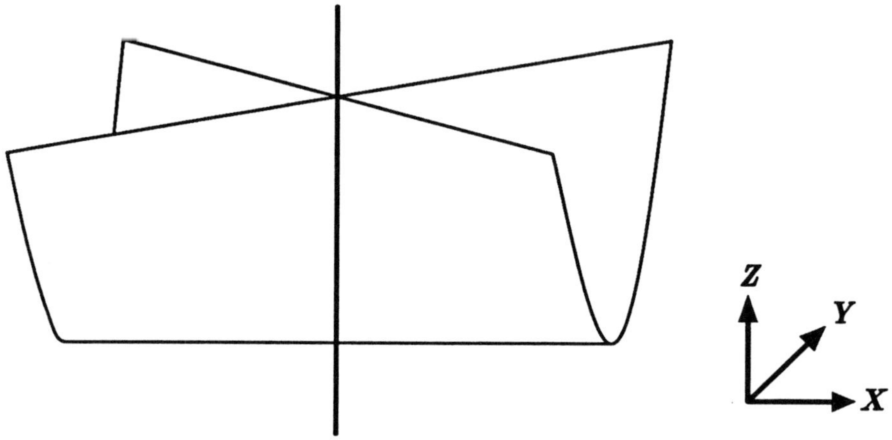

Figure 4.7: The Cartan Umbrella $y^2 = zx^2$

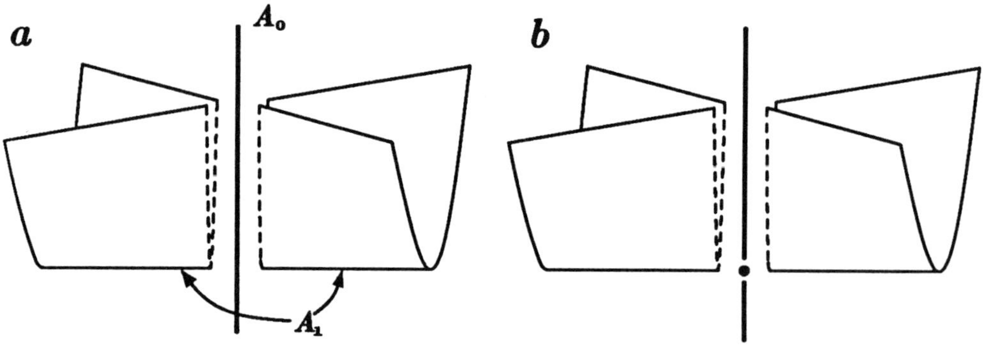

Figure 4.8: Stratification of the Cartan Umbrella. (a) Stratification which is not Whitney regular, (b) Whitney regular stratification

criteria described above, so that the stratification he described in [Wh57] was not a Whitney stratification. It was soon shown that every algebraic set also has a Whitney regular stratification, and later Wall [Wal] showed that an algebraic set has a minimal Whitney stratification which is canonical. In this thesis we will not be using this stratification of singular varieties, but instead we will simplify the problem by perturbing the polynomials so that all intersections are non-singular. We next show that the stratification into sign-invariant sets of a collection of polynomials in general position satisfies both the frontier condition and the Whitney conditions.

First we define $R^- = (-\infty, 0)$, $R^+ = (0, \infty)$, and let $\underline{\underline{\Re}}$ denote the set $\{R^-, \{0\}, R^+\}$. Then $\underline{\underline{\Re}}$ is a Whitney regular stratification of the real line. It is simple to show that a product of Whitney stratifications is also a Whitney stratification, (see [GWDL] (1.2) page 12), so that $\underline{\underline{\Re}}^n = (\underline{\underline{\Re}})^n$ is a Whitney stratification of $\Re^n$.

Definition. Let $F : \Re^r \to \Re^n$ be a polynomial map. A *sign sequence* σ is an element of $\underline{\underline{\Re}}^n$, so each σ is a manifold. Each set $F^{-1}(\sigma)$ is called a *sign-invariant set* of F. The map F is said to *define* the semi-algebraic set $S \subset \Re^r$ if S can be written as a union of sign-invariant sets of F.

We now generalize the transversality notation and write $F \pitchfork \underline{\underline{A}}$ where $\underline{\underline{A}}$ is a collection of manifolds, and take this to mean that $F \pitchfork U$ for every $U \in \underline{\underline{A}}$. Now we can show that almost any perturbation in the constant terms of all the polynomials that define S causes the sign-invariant sets to be a Whitney stratification of $\Re^r$.

General position lemma #1.
Let $F : \Re^r \to \Re^n$ be a polynomial map, $\epsilon \in \Re^n$ a parameter. Define $F_\epsilon : \Re^r \to \Re^n$ as $F_\epsilon(x) = F(x) + \epsilon$, then for almost all $\epsilon \in \Re^n$, $F_\epsilon \pitchfork \underline{\underline{\Re}}^n$, and $F_\epsilon^{-1}(\underline{\underline{\Re}}^n)$ is a Whitney stratification of $\Re^r$.

Proof. Let $\Phi : \Re^r \times \Re^n \to \Re^n$ be the map $\Phi(x, \epsilon) = F(x) + \epsilon$, then Φ is a submersion, and therefore transversal to every $\sigma \in \underline{\underline{\Re}}^n$. We apply the generic map lemma to conclude that the set $\{\epsilon \in \Re^n \mid F^\epsilon \text{ is not transversal to } \sigma\}$ has measure zero in $\Re^n$ for each σ. Since there are only finitely many σ, and the union of a finite number of measure zero sets has measure zero, $F_\epsilon \pitchfork \underline{\underline{\Re}}^n$ for almost all ϵ.

Now it is possible to pull back a Whitney stratification through a transversal map (see e.g. [GWDL]). Since $\underline{\underline{\Re}}^n$ is a Whitney stratification of $\Re^n$, and if $F_\epsilon \pitchfork \underline{\underline{\Re}}^n$ it follows that $F_\epsilon^{-1}(\underline{\underline{\Re}}^n)$ is a Whitney stratification of $\Re^r$. The elements of this stratification are the sign-invariant sets. □

Notice that if $F : \Re^r \rightarrow \Re^n$ is a polynomial map that defines the semi-algebraic set S, and $F \pitchfork \underline{\underline{\Re}}^n$, then the sign invariant sets contained in S constitute a Whitney stratification of S. In future we will assume that S can be stratified this way, and that $\underline{\underline{S}}$ denotes the stratification into sign-invariant sets.

Lemma 4.3.1 *If $F : \Re^r \rightarrow \Re^n$ defines S and $F \pitchfork \underline{\underline{\Re}}^n$ then $F^{-1}(\underline{\underline{\Re}}^n)$ satisfies the frontier condition.*

Proof We must show that the boundary of a sign-invariant set $U = F^{-1}(\sigma)$ is a union of sign-invariant sets of lower dimension. Since F is continuous, we have $\overline{U} \subset F^{-1}(\overline{\sigma})$. If ∂U denotes the boundary of U, we have $\partial U \subset F^{-1}(\partial\sigma)$. Now $\partial\sigma$ is the union of all sets σ^j, where each σ^j is formed from σ by changing one or more σ_i from R^- or R^+ to $\{0\}$. Thus ∂U is contained in the union of all strata $F^{-1}(\sigma^j)$. Since the codimension of a sign-invariant set is equal to the number of polynomials that are zero on that set, all the $F^{-1}(\sigma^j)$ have lower dimension than U.

Conversely, suppose we have a point p in some $F^{-1}(\sigma^j)$. By our general position assumption, the polynomials that are zero at p have independent differentials at p. Thus we can make an arbitrarily small displacement from p to a point q in some direction which changes only the values of those polynomials that are zero in $F^{-1}(\sigma^j)$ but non-zero in U. We can choose this displacement so that $F(q) \in \sigma$, i.e. $q \in U$. Since p has arbitrarily close neighbors in U, we have $p \in \partial U$. Thus ∂U equals the union of the $F^{-1}(\sigma^j)$. □

Example 4.6 We consider the closed cylindrical solid shown in figure 4.9. Its defining polynomials are $f_1 = x^2 + z^2 - 1$, $f_2 = y + 1$, $f_3 = y - 1$, and the set is defined by

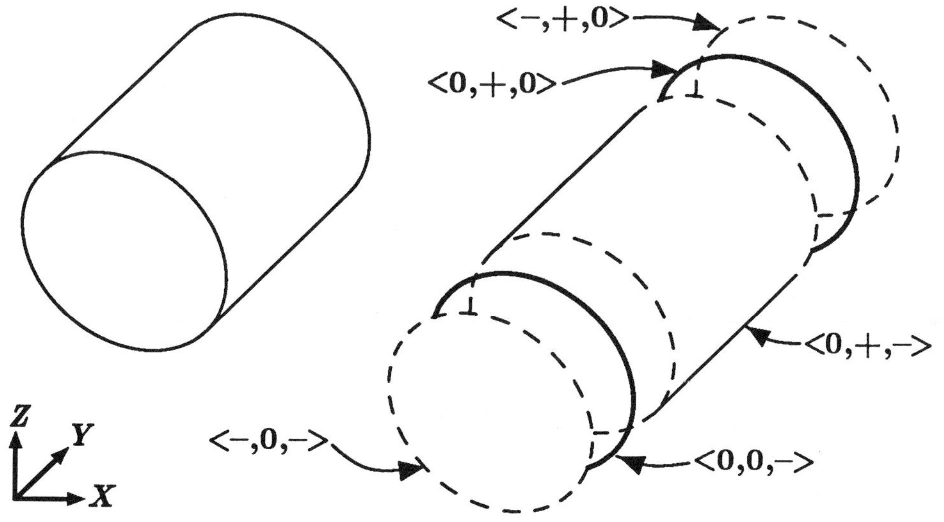

Figure 4.9: Cylinder of example 4.6 and its stratification into sign-invariant sets. Surfaces have thin outline, curves are thick. The interior stratum (sign $< -,+,- >$) is not shown.

$$(f_1 \leq 0) \wedge (f_2 \geq 0) \wedge (f_3 \leq 0) \tag{4.6}$$

As shown in the figure, the set can be broken up into smooth strata, each of which is a sign-invariant set. The sets are indexed by a triple of signs of the form $< \text{sign}(f_1), \text{sign}(f_2), \text{sign}(f_3) >$, indicating the signs of the three polynomials within that set. In this case the polynomial map (f_1, f_2, f_3) is transversal to $\underline{\underline{\Re}}^3$ so that this stratification is Whitney regular. Note that the codimension of each stratum equals the number of polynomials that are zero in its sign triple. This holds in higher dimensions as well.

Once we have $S \subset \Re^r$ represented as a stratification, for any map $f : \Re^r \to \Re^k$, we can define the critical set of $f|_{\underline{\underline{S}}}$ as the union of the critical sets of all $f|_U$ where $U \in \underline{\underline{S}}$. Critical values of $f|_{\underline{\underline{S}}}$ are the images of critical points, and all other points in $\Re^k$ are regular values. Since there are only finitely many strata, the set of critical values of $f|_{\underline{\underline{S}}}$ has measure zero in $\Re^k$.

In particular, if f is real-valued and S is semi-algebraic, $f|_{\underline{\underline{S}}}$ has a finite number of critical values. A crucial property of stratifications is that the topological type of sections $f|_{\underline{\underline{S}}}^{-1}(v)$ changes only at critical values of $f|_{\underline{\underline{S}}}$.

Theorem 4.3.2 *(Isotopy Theorem) Let $f : \Re^r \to \Re$ be a smooth mapping, and let S be a compact subset of $\Re^r$ with Whitney stratification $\underline{\underline{S}}$. If (c, d) is an open interval in $\Re$ containing no critical values of $f|_{\underline{\underline{S}}}$, then all sections $f|_S^{-1}(v)$ with $v \in (c, d)$ are isotopic, and $f|_S^{-1}(c, d)$ is homeomorphic to $f|_S^{-1}(v) \times (c, d)$.*

This result was first proved by Thom [Th69], or can be inferred from [GWDL] section II, theorem 5.2. The proof proceeds by defining a smooth vector field w_x on S which is compatible with the stratification $\underline{\underline{S}}$. A vector field w_x is said to be *compatible* with $\underline{\underline{S}}$ if for any $U \in \underline{\underline{S}}$ and any $x \in U$ then $w_x \in T_x U$. The field w_x also has the property that $df_x(w_x)$ is the unit vector field on $\Re$. The vector field w_x is locally integrable, and the flow it defines gives the isotopy between sections. Let $\phi_x(t)$ be an integral curve of this field in $\Re^r$ such that $\phi_x(0) = x$, and

$$\frac{\partial \phi_x}{\partial t} = w_x \tag{4.7}$$

from which it follows that $f(\phi_x(t)) = f(x) + t$. Each such curve is defined on an open interval $(t_x^-, t_x^+) \subset \Re$, and the domain of the flow $\phi(x, t) = \phi_x(t)$ is an open subset of $\Re^r \times \Re$. From now on we will use the slice notation $S|^B = f|_S^{-1}(B)$ where B is either a real value or an interval in $\Re$.

We wish to extend ϕ so that we can define a continuous deformation retraction from $S|^{(-\infty,d)}$ to $S|^{(-\infty,c]}$. If c is a critical value of $f|_{\underline{\underline{S}}}$, then there is a critical point x_c in its domain at which the flow is undefined (there is exactly one such point if f is in general position). Now each ϕ_x with $x \in S|^{(c,d)}$ has $t_x^- \leq c - f(x)$ because w_x is non-vanishing throughout $S|^{(c,d)}$. If $t_x^- = c - f(x)$, then we extend ϕ_x to a new curve $\bar{\phi}_x : [t_x^-, t_x^+) \to \Re^r$ as follows:

Let (t_n) be a sequence of points in $(c - f(x), d - f(x))$ which converges to $c - f(x)$. By compactness of S, $(\phi_x(t_n))$ has a convergent subsequence. The limit of this sequence must be the point x_c, since any other point has smooth flow defined on it. We set $\bar{\phi}_x(c - f(x)) = x_c$ for all ϕ_x such that $t_x^- = c - f(x)$.

Claim. Let $D = \{(x, t) \mid t \in [c - f(x), d - f(x))\}$. The extension of flow $\bar{\phi} : D \to \Re^r$ defined as $\bar{\phi}(x, t) = \bar{\phi}_x(t)$ (appropriately restricted) is continuous.

Proof. Since x_c is the only new point in the image of $\bar{\phi}$, it suffices to show that if (x_i, t_i) is any sequence converging to $(x, t) \in \bar{\phi}^{-1}(x_c)$, then $\bar{\phi}(x_i, t_i)$ converges to x_c. Infinitely many $\bar{\phi}(x_i, t_i)$ must lie in some compact slice $S|^{[c,c+\epsilon]}$ and so must have a limit point. Suppose there exists a limit point $x_0 \neq x_c$, first we notice that $f(x_0) = c$. Let $U_0 \subset S|^c$ be an open neighborhood of x_0 (in $S|^c$) whose closure does not contain x_c. Now ϕ induces a homeomorphism from $(S|^c - \{x_c\}) \times [c, d)$ to $S|^{[c,d)}$, and so $T = \phi(U_0, [0, d-c))$ is a tube whose boundary separates $S|^{[c,d)}$. Similarly, $\partial T \times \Re$ is a closed set which separates the domain of $\bar{\phi}$ into open sets, one containing (x, t) and the other containing $\bar{\phi}^{-1}(T)$. Since (x_i, t_i) converges to (x, t), $\bar{\phi}^{-1}(T)$ contains only finitely many points of (x_i, t_i). Therefore T is an open neighborhood of x_0 containing only finitely many points of the sequence $\bar{\phi}(x_i, t_i)$, so x_0 cannot be a limit point of this sequence. □

Retraction Lemma. Let $(c, d) \subset \Re$ be an interval containing no critical values of $f|_{\underline{\underline{S}}}$, with c possibly a non-degenerate critical value. Then $S|^{(-\infty,c]}$ is a deformation retract of $S|^{(-\infty,d)}$.

Proof.

We define a retraction $r : S|^{(-\infty,d)} \times I \to S|^{(-\infty,d)}$ as:

$$r(x, t) = \begin{cases} \bar{\phi}(x, (c - f(x))t) & \text{if } f(x) \in [c, d); \\ x & \text{otherwise.} \end{cases} \tag{4.8}$$

then r is defined throughout $S|^{(-\infty,d)} \times I$, and is continuous, by continuity of $\bar{\phi}$. We have $r(x,0) = x$, $r(x,t) = x$ for $f(x) \leq c$, and $r(x,1) \subset S|^{(-\infty,c]}$ for all x. Thus r is the desired retraction. □

4.4 The Silhouette

In the construction of the silhouette, we will need to compute the extremal points in some stratum $U \in \underline{\underline{S}}$ with respect to some coordinate, say x_i. These extremal points are the critical points of the projection map on the ith coordinate. Let $\pi_i : \Re^r \to \Re$ denote projection on the ith coordinate, and we generalize this notation to define $\pi_{i_1,\ldots,i_k} : \Re^r \to \Re^k$ as projection on the coordinates i_1 through i_k. Then the extremal points of U with respect to x_i are exactly the critical points of the map $\pi_i|_U$. In our case, each stratum U is an open subset of the kernel of some map $f : \Re^r \to \Re^n$, and in this case there is a simple test for critical points of any map $g|_U$:

Lemma 4.4.1 *Suppose that U is an open subset of the kernel of some map $f : \Re^r \to \Re^n$, and that $f \pitchfork \{0\}$. Let $g : \Re^r \to \Re^m$ be a map, then $x \in U$ is a critical point of $g|_U$ iff the matrix*

$$d(f,g)_x = \begin{pmatrix} \frac{\partial f_1}{\partial x_1}(x) & \cdots & \frac{\partial f_1}{\partial x_r}(x) \\ \vdots & & \vdots \\ \frac{\partial f_n}{\partial x_1}(x) & \cdots & \frac{\partial f_n}{\partial x_r}(x) \\ \frac{\partial g_1}{\partial x_1}(x) & \cdots & \frac{\partial g_1}{\partial x_r}(x) \\ \vdots & & \vdots \\ \frac{\partial g_m}{\partial x_1}(x) & \cdots & \frac{\partial g_m}{\partial x_r}(x) \end{pmatrix} \tag{4.9}$$

has rank less than $m + n$ at x.

Proof. We note that the first n rows of $d(f,g)_x$ are the rows of df_x, while the second m rows are the rows of dg_x. The point x is a critical point of $g|_U$ iff $d(g|_U)_x$ is not surjective, i.e. iff $\dim(dg_x(T_xU)) < m$. This will be true iff there is a non-zero vector $u \in T_xU$ such that $dg_x(u) = 0$. But T_xU is just $\ker(df_x)$ so u must be orthogonal to all the rows of df_x and all the rows of dg_x. Such a u exists iff $d(f,g)_x$ has rank less than $m + n$ at x. □

Corollary 4.4.2 *The critical set $\Sigma(g|_U)$ is a closed subset of U since it is defined by the vanishing of a finite number of subdeterminants of $d(f,g)_x$. If U is semi-algebraic, g a polynomial map, then $\Sigma(g|_U)$ is semi-algebraic.*

Example 4.6 Let f be the polynomial $f(x,y,z) = x^2 + y^2 + z^2 - 1$. Then $U = \ker(f)$ is the unit 2-sphere in $\Re^3$ as in example 4.2. Suppose $g = \pi_1$ is the function of interest. Then the critical points of $g|_U$ are the points on U where the matrix

$$d(f,g) = \begin{pmatrix} 2x & 2y & 2z \\ 1 & 0 & 0 \end{pmatrix} \tag{4.10}$$

is singular. The matrix has rank less than 2 iff $y = z = 0$ and the only points on the manifold satisfying this condition are $(-1,0,0)$ and $(1,0,0)$. These are the points where the function $g|_U$ attains its extremal values of -1 and 1.

Lemma 4.4.3 *(Slice Lemma.) Let $M \subset \Re^r$ be a manifold which is the kernel of some map $f : \Re^r \to \Re^n$, with $f \pitchfork \{0\}$. Then the critical set $\Sigma(\pi_{12}|_M)$ is the union of the critical set $\Sigma(\pi_1|_M)$ and all critical sets $\Sigma(\pi_2|_N)$, where N is a slice of the form $N = \pi_1|_M^{-1}(v) - \Sigma(\pi_1|_M)$ for some $v \in \Re$.*

Proof. First we form a matrix of the form (4.9) whose rows are the rows of the jacobian of f followed by π_1, followed by π_2. Then $\Sigma(\pi_{12}|_M)$ is the set of points in M where the matrix is singular. This set can be broken into two subsets corresponding to the two ways that the matrix can be singular:

(i) the first $n+1$ rows are dependent, which is true iff x is a critical point of $\pi_1|_M$.

(ii) the last row is dependent on the first $n+1$, which is true iff x is a critical point of $\pi_2|_N$ for some slice $N = \pi_1|_M^{-1}(v)$ of M. □

Example 4.8 Suppose that M is again the sphere defined by $f(x,y,z) = x^2+y^2+z^2-1$. For the function $f = \pi_{12}|_M$ (the same function as in example 4.2), the matrix (4.9) has the form

$$d(f,g) = \begin{pmatrix} 2x & 2y & 2z \\ 1 & 0 & 0 \\ 0 & 1 & 0 \end{pmatrix} \tag{4.11}$$

and this matrix is singular iff $z = 0$. Thus $C = \Sigma(\pi_{12}|_M)$ is the unit circle in the x-y plane, as we saw in figure 4.4. The critical points of $\pi_1|_M$ are the two points $(-1,0,0)$ and $(1,0,0)$. Each set $\pi_1^{-1}(v)$ is a circle which is a slice through M at $x = v$. The critical points of $\pi_2|_N$ on each slice $N = \pi_1^{-1}(v) - \Sigma(\pi_1|_M)$ are the extremal points in y of the slice, and satisfy the constraint $z = 0$. The union of the critical points of all such slices together with the critical points of $\pi_1|_M$ gives the curve C.

Lemma 4.4.4 *Let $f : \Re^r \rightarrow \Re^k$ be a map, and assume S is closed with Whitney stratification $\underline{\underline{S}}$. Then the critical set $\Sigma(f|_{\underline{\underline{S}}})$ is closed.*

Proof. Let $U \in \underline{\underline{S}}$ be a stratum, and let (x_i) be a sequence of points in $\Sigma(f|_U)$ which converges to x. If x lies in U, then it must also be in $\Sigma(f|_U)$ since $\Sigma(f|_U)$ is a closed subset of U. If $x \notin U$, then it must be in some stratum $V \in \underline{\underline{S}}$ adherent to U. Passing to a subsequence if necessary, we must have $(T_{x_i}U) \rightarrow \tau$ for some τ. By Whitney's condition (A), we have $T_xV \subset \tau$. By continuity of the differential, we must have $df_x(\tau) \neq \Re^k$, and therefore $df_x(T_xV) \neq \Re^k$, so that x is a critical point of $f|_V$. It follows that $\Sigma(f|_{\underline{\underline{S}}}) \cap \overline{U}$ is closed for every stratum $U \in \underline{\underline{S}}$, and therefore $\Sigma(f|_{\underline{\underline{S}}})$ is closed in S. □

Later we will need to be able to choose a projection function a such that $\Sigma(a|_U)$ is a 1-manifold for any $U \in \underline{\underline{S}}$. This will be the case if the map a is sufficiently "generic". In order to formalize this and prove the result, we must introduce jets. Our development is rather terse, a more complete one is given in [GG].

Let M and N be manifolds of dimension m and n respectively. Let $f : M \rightarrow N$ be a smooth map and $x \in M$. We define an equivalence relation $\sim_k$ on pairs (f, x) such that $(f, x) \sim_k (f', x')$ iff $x = x'$, $f(x) = f'(x')$ and in local coordinates in M and N centered at x and $f(x)$ respectively, the partial derivatives at x of f and f' up to order k are equal.

Definition Let $[f,x]_k$ denote the equivalence class of (f,x) under $\sim_k$, then $[f,x]_k$ is called a *k-jet of mappings* from M to N with *source* x and *target* $f(x)$. The set of all k-jets from M to N is denoted $J^k(M,N)$.

$J^k(M,N)$ can be given the structure of a manifold, in fact a fibre bundle over $M \times N$. We will be interested in the special case $N = \Re^n$, and for this case, $J^k(M,N)$ is a vector bundle over $M \times N$. For example, $J^1(M,\Re^n)$ is a vector bundle with fibre $\Re^{mn}$, since this fibre is isomorphic to the space of linear transformations from $\Re^m$ to $\Re^n$.

For each $f : M \to N$, there is a canonically defined mapping $j^k f : M \to J^k(M,N)$ such that $j^k f(x) = [f,x]_k$, which is called the k-jet of f. The k-jet $j^k f(x)$ may be thought of as an invariant k^{th}-order Taylor expansion of f at x. For example, $j^1 f(x)$ is the equivalence class of functions that have the same value and differential as f at x.

Not surprisingly, the notion of criticality can be formulated as a condition on 1-jets. We will assume from now on that $m \geq n$, we define for $s = 1, \ldots n$

$$\Sigma^s = \{[f,x]_1 \in J^1(M,N) \mid \text{corank}(df_x) = s\} \tag{4.12}$$

then the Σ^s contain in some sense, the critical sets of all maps between M and N. If $\Sigma^s(f)$ denotes the subset of $\Sigma(f)$ where df has corank s, then we can pull down the singular locus for particular maps, i.e.

$$\Sigma^s(f) = (j^1 f)^{-1}(\Sigma^s) \tag{4.13}$$

Importantly, Σ^s is a submanifold of $J^1(M,N)$, in fact a subfibre bundle over $M \times N$. Its fibre is $L^s(\Re^m,\Re^n)$, which is the set of linear transformations from $\Re^m$ to $\Re^n$ which have corank s. $L^s(\Re^m,\Re^n)$ is itself a submanifold of the space $L(\Re^m,\Re^n)$ of all linear transformations from $\Re^m$ to $\Re^n$, and it is readily shown that $\text{codim}(L^s(\Re^m,\Re^n)) = (m-n)s + s^2$. This implies that $\text{codim}(\Sigma^s) = (m-n)s + s^2$ and using the codimension expression for transversal maps, the following lemma is immediate:

Lemma 4.4.5 *Let* $f : M \to N$ *be a smooth map with* $\dim(M) = m$, $\dim(N) = n$ *and* $m \geq n$. *If* $j^1 f \pitchfork \Sigma^s$, *then* $\Sigma^s(f) = (j^1 f)^{-1}(\Sigma^s)$ *is a submanifold of* M *of codimension* $(m-n)s + s^2$.

The transversality condition of the previous lemma is given a special name. We say a map $f : M \to N$ is *one-generic* if $j^1 f \pitchfork \Sigma^s$ for all s. If we can satisfy this condition, we will greatly simplify the form of the critical sets $\Sigma(f)$. So next we show:

Lemma 4.4.6 *Let $a : \Re^r \to \Re^n$ be a linear map, $M \subset \Re^r$ a manifold of dimension $m \geq n$. Then a can be specified by rn real values, and for almost all $a \in \Re^{rn}$, $a|_M$ is one-generic.*

Proof. We first define a parametric map $\Phi : M \times \Re^{rn} \to J^1(M, \Re^n)$ as $\Phi(x, a) = j^1 a|_M$. If $\Phi \pitchfork \Sigma^s$ for some s, then $j^1 a|_M \pitchfork \Sigma^s$ for almost all specializations a by the generic map lemma. If we can show $\Phi \pitchfork \Sigma^s$ for all s, then we are done since there are only finitely many s. At any "point" $p = (x, a) \in \Phi^{-1}(\Sigma^s)$ we have

$$d\Phi_p(T_p(M \times \Re^{rn})) = D_x \times \Re^{mn} \tag{4.14}$$

where D_x is some subspace of the space $T_x M \times \Re^n$. Since Σ^s is a fibre bundle over $M \times \Re^n$, the tangent space to Σ^s at the point $\Phi(p)$ is

$$T_{\Phi(p)}\Sigma^s = T_x M \times \Re^n \times T_a L^s(\Re^m, \Re^n) \tag{4.15}$$

and the sum of these two spaces is $T_x M \times \Re^n \times \Re^{mn}$, which is the tangent space to $J^1(M, \Re^n)$ at $\Phi(p)$. Thus $\Phi \pitchfork \Sigma^s$ for any s and we are done. $\square$

General position lemma #2.
Let $a : \Re^r \to \Re^2$, and $M \subset \Re^r$ a manifold of $\dim(M) \geq 1$. Then for almost all $a \in \Re^{2r}$, the critical set $\Sigma(a|_M)$ is a one-dimensional manifold.

Proof. If $m = \dim(M) = 1$, then $\Sigma(a|_M) = M$ and the result is trivial. Otherwise, apply the previous lemma so that $j^1(a|_M) \pitchfork \Sigma^s$. Maps into $\Re^2$ can have corank at most 2, so $\Sigma(a|_M) = \Sigma^1(a|_M) + \Sigma^2(a|_M)$. Now $\Sigma^1(a|_M)$ is a manifold of codimension $(m-1)$ in M, so it is one-dimensional. But $\Sigma^2(a|_M)$ has codimension $2m$, i.e. it is empty, so $\Sigma(a|_M) = \Sigma^1(a|_M)$ and is a one-manifold. $\square$

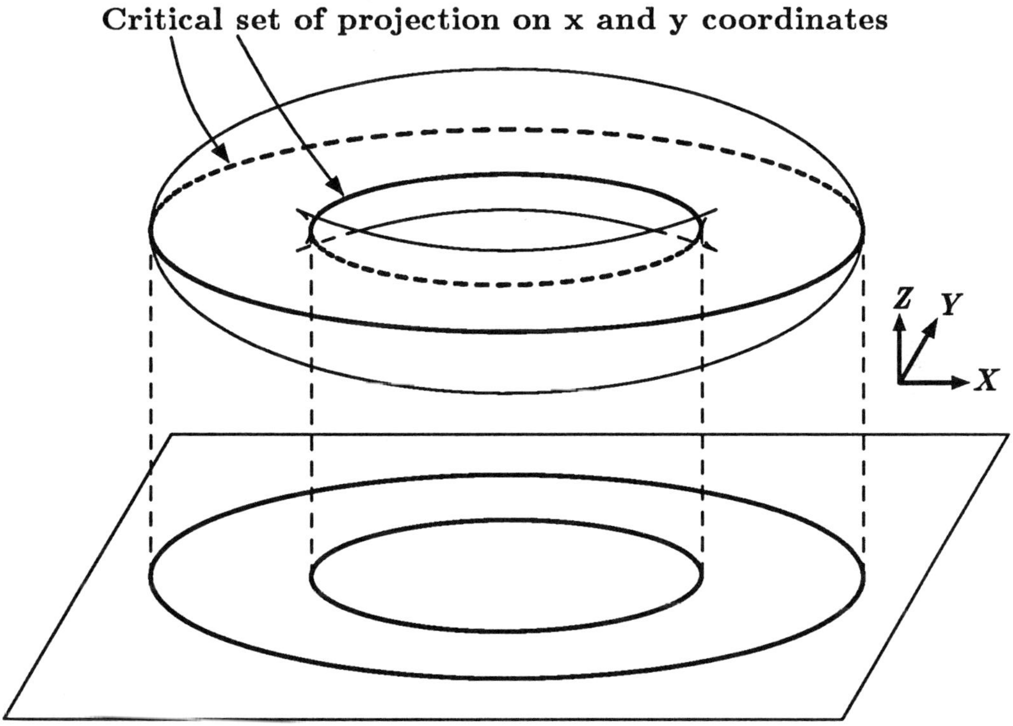

Figure 4.10: The silhouette of the torus

Example 4.9 The silhouette of the torus. Consider a torus with its main axis normal to the x-y-plane as shown in figure 4.10. For projection down on the x-y-plane, the critical set consists of two circles on the inner and outer perimeter of the torus. From the figure, the reader can see the origin of the term "silhouette" although, in our case, the silhouette of the torus is the set of points on the torus itself which project onto the x-y-plane to what would usually be called the silhouette.

We saw earlier that the silhouette $C = \Sigma(\pi_{12}|_M)$ contains the critical set $\Sigma(\pi_1|_M)$. Now we know that C is actually a submanifold of M (assuming π_{12} is one-generic), and from this it follows that critical points of $\pi_1|_M$ are also critical points of $\pi_1|_C$. It is natural to ask under what conditions the reverse is true, i.e. under what conditions are critical points of $\pi_1|_C$ also critical points of $\pi_1|_M$? This brings us naturally to an examination of higher-order singularities.

Definition Let $f : M \rightarrow \Re^2$ be a one-generic map with $C = \Sigma(f)$ its critical set. A point $x \in C$ is called a *cusp point* if $df_x(T_xC) = 0$, otherwise x is called a *fold point.*

Lemma 4.4.7 *Let $\pi_{12}|_M$ be one-generic, $C = \Sigma(\pi_{12}|_M)$ its critical set. If x is a critical point of $\pi_1|_C$ and a fold point, then it is a critical point of $\pi_1|_M$.*

Proof. Since x is fold point, $d(\pi_{12})_x(T_xC)$ is one-dimensional, and we must have $d(\pi_{12})_x(T_xM) \subset d(\pi_{12})_x(T_xC)$, otherwise $d(\pi_{12})_x$ would be surjective, contradicting the fact that x is a critical point of $\pi_{12}|_M$ (this is not true for cusps, where $d(\pi_{12})_x(T_xC) = 0$, and $d(\pi_{12})_x(T_xM)$ can be any one-dimensional space). But $d(\pi_1)_x(T_xC) = 0$, and therefore $d(\pi_1)_x(T_xM) = 0$, so that x is a critical point of $\pi_1|_M$. □

If the condition described above were always true, i.e. if critical points of $\pi_1|_C$ were also critical points of $\pi_1|_M$, it would considerably simplify the roadmap algorithms to be described later. Unfortunately, cusps occur generically (in stable views of the torus for example), so there is not much that can be done about them. It is still useful to recognise critical points as fold points in non worst case examples though, as we shall see later.

4.4.1 Tubes Around Varieties

We can reduce the problem of computing critical points on varieties of arbitrary dimension to finding critical points on certain hypersurfaces. Suppose we wish to find the critical points of the map $a : \Re^r \rightarrow \Re$ restricted to the non-singular algebraic set M defined by $f_1 = 0, \ldots f_n = 0$. We define a new polynomial

$$g = \sum_i f_i^2 \tag{4.16}$$

A similar construction was used by both Thom [Th65] and Milnor [Mil64] in their bounds for Betti numbers of varieties. We next prove a local geometric result which is closely related to the global bounds in the above papers. The result allows us to find critical points of arbitrary manifolds by considering only hypersurfaces. We show that if p is a non-degenerate critical point of $a|_M$, then for small enough ϵ, there is a critical point of $a|_{g^{-1}(\epsilon)}$ "near" p. A non-degenerate critical point of $a|_M$ is a critical point where the Hessian (matrix of second partial derivatives) is non-singular. That is, the Hessian H of some function $f : M \rightarrow \Re$ is defined by $H_{ij} = \frac{\partial^2 f}{\partial x_i \partial x_j}$ in local coordinates. A generically chosen map into $\Re$ has only non-degenerate critical points.

Let $p \in M$ be a non-degenerate critical point of $a|_M$, and let $v = a(p)$. We show that any ball $B_\rho = \{q : d(p - q) < \rho\}$ about p contains a critical point of $a|_{g^{-1}(\epsilon)}$ for all sufficiently small ϵ. ($d(\cdot)$ is the usual euclidean norm on $\Re^r$)

Let $N = M \cap \overline{B}_\rho$ denote the intersection of M with the closure of B_ρ. N is contained in a compact "ϵ-neighborhood" K, where $K = \overline{B}_\rho \cap g^{-1}[0, \epsilon]$. K can also be defined as follows: let $F : \Re^r \rightarrow \Re^2$ be the map $F(q) = (g(q), d(p - q))$, then $K = F^{-1}([0, \epsilon] \times [0, \rho])$.

Lemma 4.4.8 *N is a deformation retract of K.*

Proof Define $W = (0, \epsilon] \times (0, \rho]$, then $F^{-1}(W)$ equals $K - N$. We assume that g, when restricted to either B_ρ or its boundary, has no critical points in the range $(0, \epsilon]$. Since g has only finitely many critical values when restricted to either stratum (its critical points in each case are defined by an algebraic equation of the form (4.9)), this is easily satisfied by choosing ϵ small enough. We can then use g to lift a vector field (with value -1) on $(0, \epsilon]$ to the set $K - N$, and this lifting is compatible with the stratum B_ρ and its boundary. Call the lifted vector field η_q, then $dg_q(\eta_q) = -1$ for any $q \in K - N$. The field η_q defines a continuous flow on $K - N$, and the limit points of flow lines lie in N. The flow extends to a continuous function on K which gives us a deformation retraction of K into N. Thus for small enough ϵ, K and N have the same homotopy type. □

In the following lemmas, we use the slice notation $A|^{\leq u}$ to mean the intersection of the set A with $a^{-1}(-\infty, u]$, e.g. $N|^{\leq v+\delta} = N \cap a^{-1}(-\infty, v+\delta]$.

Lemma 4.4.9 *The set $N|^{\leq v-\delta}$ is a deformation retract of $K|^{\leq v-\delta}$, and $N|^{\leq v+\delta}$ is a deformation retract of $K|^{\leq v+\delta}$.*

Proof The proof follows that of lemma 4.4.8. We stratify the set $\overline{B}_\rho|^{\leq v-\delta}$ using the map $F : \Re^r \to \Re^2$ where $F(q) = (d(p-q), a(q))$. We apply this map to the stratified set $\underline{\underline{W}} = \{(0, \rho), \{\rho\}\} \times \{(-\infty, v-\delta), \{v-\delta\}\}$, and by the first general position lemma (this chapter), for almost all choices of ρ, δ the preimage $F^{-1}(\underline{\underline{W}})$ is a Whitney regular stratification of $\overline{B}_\rho|^{\leq v-\delta}$. Now for small enough ϵ, the map g has no critical values in the range $(0, \epsilon]$ when restricted to any of the strata of $F^{-1}(\underline{\underline{W}})$. Thus we can use g to lift a constant vector field on $(0, \epsilon]$ up to $F^{-1}(\underline{\underline{W}}) \cap g^{-1}(0, \epsilon]$ which is compatible with the stratification. Now $F^{-1}(W) \cap g^{-1}(0, \epsilon]$ equals $(K - N)|^{\leq v-\delta}$, and the lifted vector field extends to give us a retraction of $K|^{\leq v-\delta}$ onto $N|^{\leq v-\delta}$. The proof for sections by $a^{-1}(-\infty, v+\delta]$ is identical. □

Lemma 4.4.10 *For all sufficiently small δ, the homotopy types of the sets $N|^{\leq v-\delta}$ and $N|^{\leq v+\delta}$ are different.*

Proof N has a natural, Whitney regular stratification into subsets $N_1 = N \cap B_\rho$ and $N_2 = N \cap \partial B_\rho$, where ∂B_ρ is the boundary of B_ρ. We assume that $a|_{N_2}$ does not have v as a critical value. This will be true for generic choices of a and fixed N_1 and N_2. Alternatively, we could invoke the local coordinate lemma of Morse [Mil63] page 6, that the value of a on the manifold M can be written locally as

$$a = v + u_1^2 + \cdots + u_\lambda^2 - u_{\lambda+1}^2 - \cdots - u_{r-n}^2$$

for suitable local coordinates $u_1, \ldots, u_{r-n}$ on M centered at p. The set N_2 is defined as the set of points where some weighted sum w of the u_i^2 is constant, while $a - v$ is defined as a difference of u_i^2's. If $a = v$ at a point $q \in N_2$, both some u_i for $i \leq \lambda$ and some u_j for $j > \lambda$ must be non-zero. Here the functions w and $a - v$ have independent differentials, so q is not a critical point of $a|_{N_2}$.

Having shown that v is not a critical value of $a|_{N_2}$ we now choose δ so that there are no critical values of $a|_{N_2}$ in the range $[v - \delta, v + \delta]$. By the main theorem of Morse theory [Mil63] page 14, the set $N|^{\leq v+\delta}$ has the homotopy type of $N|^{\leq v-\delta}$ with a λ-cell adjoined. Here λ is the *index* of the critical point (it is the same λ as in the Morse lemma above) We will not define the index here, since all we need to know is that the homotopy types of the two sets are *different*. Thus $N|^{\leq v-\delta}$ *cannot* be a deformation retract of $N|^{\leq v+\delta}$ if δ is small enough. □

Lemma 4.4.11 *For all sufficiently small δ and $\epsilon(\delta)$, if $H = g^{-1}(\epsilon)$ and there are no critical points of $a|_H$ in the ball B_ρ, then $K|^{\leq v-\delta}$ is a deformation retract of $K|^{\leq v+\delta}$.*

Proof In this case we define a map $F : \Re^r \to \Re^2$ as $F(q) = (g(q), d(p-q))$. We then stratify the set $W = (0, \epsilon] \times (0, \rho]$ into $\underline{\underline{W}} = \{(0, \epsilon), \{\epsilon\}\} \times \{(0, \rho), \{\rho\}\}$ which is Whitney regular. Applying the result of the first general position lemma, for almost all pairs (ϵ, ρ), F is transversal to $\underline{\underline{W}}$, and so the preimage $F^{-1}(\underline{\underline{W}})$ is a Whitney regular stratification. Adding the limit points where $g = 0$ to the strata having $g < \epsilon$ gives a Whitney regular stratification of K. We denote this as $\underline{\underline{K}}$.

$\underline{\underline{K}}$ has four strata, and by the hypothesis of this lemma, one of them $K_1 = H \cap B_\rho$ contains no critical points of the restriction of a. The "interior" stratum K_2 defined by $\{q | g(q) < \epsilon,\ d(p-q) < \rho\}$ has no critical points since it is an open subset of $\Re^r$. The stratum K_3 defined by $\{q | g(q) < \epsilon,\ d(p-q) = \rho\}$ will give rise to no critical values in the range $[v-\delta, v+\delta]$ for small enough δ since it is an open subset of ∂B_ρ. (There will be two critical points on B_ρ but these give rise to extremal values $v - h_1$ and $v + h_2$. If δ is chosen between these extremals, there are no critical points in the range)

Finally, we claim that the last stratum K_4 defined by $\{q | g(q) = \epsilon,\ d(p-q) = \rho\}$ has no critical values of the restriction of a in the range $[v - \delta, v + \delta]$ for small enough ϵ. We assume that δ is chosen as in the previous lemma so that $a|_{N_2}$ has no critical values in the range $[v - \delta, v + \delta]$.

Suppose there were no ϵ_0 such that $\epsilon < \epsilon_0$ implies that $a|_{K_4}$ does not have a critical value in $[v - \delta, v + \delta]$. Then we could choose an infinite decreasing sequence $(\epsilon_i) \to 0$ which has a sequence of critical points (q_i) on K_4's. Passing to a subsequence if necessary, there is a limit point q which lies in N_2 and which must be a critical point of $a|_{N_2}$. This is because we have $da_{q_i}(T_{q_i}K_4) = 0$, and the tangent space at the limit point $T_q N_2$ must be contained in the limit of tangent spaces of the sequence $(T_{q_i}K_4)$ (the limit is taken in the grassmanian bundle over $\Re^r$). By continuity of the differential this implies $da_q(T_q N_2) = 0$, i.e. that q is a critical point of $a|_{N_2}$. Now $a(q) \in [v - \delta, v + \delta]$, but δ was chosen so that there are no such critical points, a contradiction. So our assumption that there was no ϵ_0 must be false.

So if δ is small enough, and $\epsilon(\delta)$ is then chosen sufficiently small, there are no critical values of a restricted to any of the strata in the stratification of K (assuming the hypothesis of this lemma about K_1 holds !). Thus we can use a to lift a constant vector field on $\Re$ to a field η_q on K which is compatible with the stratification $\underline{\underline{K}}$. The field is chosen to satisfy $da(\eta_q) = -1$. This field defines the desired deformation retraction from $K|^{\leq v+\delta}$ to $K|^{\leq v-\delta}$. □

Theorem 4.4.12 *If p is a non-degenerate critical point of a_M, then for any $\rho > 0$, and for all sufficiently small $\epsilon > 0$, there is a critical point q of $a|_{g^{-1}(\epsilon)}$ such that $d(p - q) < \rho$.*

Proof From lemmas 4.4.10 and 4.4.9 we know that if δ and $\epsilon(\delta)$ are small enough, then in terms of homotopy type

$$K|^{\leq v-\delta} \approx N|^{\leq v-\delta} \not\approx N|^{\leq v+\delta} \approx K|^{\leq v+\delta}$$

where $\approx$ means equivalence of homotopy type. So $K|^{\leq v-\delta} \not\approx K|^{\leq v+\delta}$. But if there is no critical point of $a|_{g^{-1}(\epsilon)}$ in the ball B_ρ, then lemma 4.4.11 implies that $K|^{\leq v-\delta} \approx K|^{\leq v+\delta}$, a contradiction. Hence there must be such a critical point, and the argument holds for all sufficiently small ϵ. □

4.5 The Roadmap

We now have the hardware necessary to construct a one dimensional semi-algebraic subset of S which we call the roadmap of S and which we denote $R(S)$. This subset serves as a skeleton for motion planning and provides candidate paths for motion between any two configurations. In order to guarantee that solutions may be found along the roadmap, it must satisfy the following condition:

Definition A subset R of a set S satisfies the *roadmap condition* if every connected component of S contains a single connected component of R.

Before describing the roadmap, we define a refinement of the stratification $\underline{\underline{S}}$ of S, to give a new Whitney stratification $\underline{\underline{S}}'$, called the *Silhouette compatible stratification* of S. This stratification includes as a subset, a stratification of the silhouette $\Sigma = \Sigma(\pi_{12}|_{\underline{\underline{S}}})$.

Let $\underline{\underline{\Sigma}} = \{\Sigma(\pi_{12}|_U) | U \in \underline{\underline{S}}\}$ be a stratification of Σ. This is not a Whitney stratification, since a boundary point of one curve may lie inside another. If however, we define $B = \{\partial C | C \in \underline{\underline{\Sigma}}\}$ as the set of such points, we can construct a finer stratification of Σ which is a Whitney stratification. Let $\underline{\underline{B}}$ be the stratification of B into one point sets, then

$$\underline{\underline{\Sigma}}' = \underline{\underline{B}} \cup \{U - B | U \in \underline{\underline{\Sigma}}\} \tag{4.17}$$

is a Whitney stratification of Σ. If now we define

$$\underline{\underline{S}}' = \underline{\underline{\Sigma}}' \cup \{U - \Sigma | U \in \underline{\underline{S}}\} \tag{4.18}$$

then $\underline{\underline{S}}'$ is a Whitney stratification of S, which includes $\underline{\underline{\Sigma}}'$ as a subset.

In what follows we will use the notation $\underline{\underline{A}} \cap V = \{U \cap V | U \in \underline{\underline{A}}\}$ and the slice notation $\underline{\underline{A}}|^B = \underline{\underline{A}} \cap \pi_1^{-1}(B)$ where B is a point or an interval in $\Re$.

4.5.1 The Basic Roadmap

Definition. Let $\underline{\underline{S}}$ be a Whitney stratification of a compact set. Then the roadmap $R_0(\underline{\underline{S}})$ is defined inductively as follows:

1. If $\dim(\underline{\underline{S}}) = 1$, then $R_0(\underline{\underline{S}}) = S$.

2. Otherwise let $\Sigma = \Sigma(\pi_{12}|_{\underline{\underline{S}}})$ be the silhouette of S and let $\underline{\underline{S}}'$ be the silhouette compatible stratification of S. Let $P_c \subset \Re$ be the (finite) set of critical values of $\pi_1|_{\underline{\underline{S}}'}$. The roadmap $R_0(\underline{\underline{S}})$ is given by

$$R_0(\underline{\underline{S}}) = \Sigma \cup \bigcup_{v \in P_c} R_0(\underline{\underline{S}}'|^v) \tag{4.19}$$

That is, the roadmap of a one dimensional set is the set itself. Otherwise the roadmap is the union of the silhouette Σ and the roadmaps of slices through $\underline{\underline{S}}'$ at critical values of $\pi_1|_{\underline{\underline{S}}'}$. The construction is well defined because if $\dim(\underline{\underline{S}}) = k$ then the slice $\underline{\underline{S}}'|^v$ has dimension at most $k-1$.

Example 4.10 The roadmap of the torus. We consider again the torus with major axis normal to the x-y-plane, as in example 4.9. As we saw in figure 4.10 the silhouette Σ of the torus is not connected. However, in figure 4.11, we see that the roadmap consists of the silhouette and in addition, the recursively computed roadmaps at slices $x = c$ through critical points of $\pi_1|_\Sigma$. In this case, there are four critical points, but for two of these, the slice consists only of a single point. For the remaining two, the slices are already one-dimensional, and consist of pairs of curves which meet at the critical points. Since they are one-dimensional, the roadmap of each slice equals the slice.

We claim that $R_0(\underline{\underline{S}})$ satisfies the roadmap condition, but to prove this we need the following inductive lemma:

Lemma 4.5.1 *If for each slice $S|^v$ through a critical value $v \in P_c$, $R_0(\underline{\underline{S}}'|^v)$ satisfies the roadmap condition, then $R_0(\underline{\underline{S}})$ satisfies the roadmap condition.*

Proof. Let $v_1, \ldots v_m$ be the critical values of $\pi_1|_{\underline{\underline{S}}'}$ arranged in ascending order. Define a new sequence of points $u_0, \ldots u_m$ as follows

$$u_i = \begin{cases} v_1 & \text{if } i = 0; \\ \frac{v_i + v_{i+1}}{2} & \text{for } i = 1, \ldots m-1; \\ v_m & \text{if } i = m. \end{cases} \tag{4.20}$$

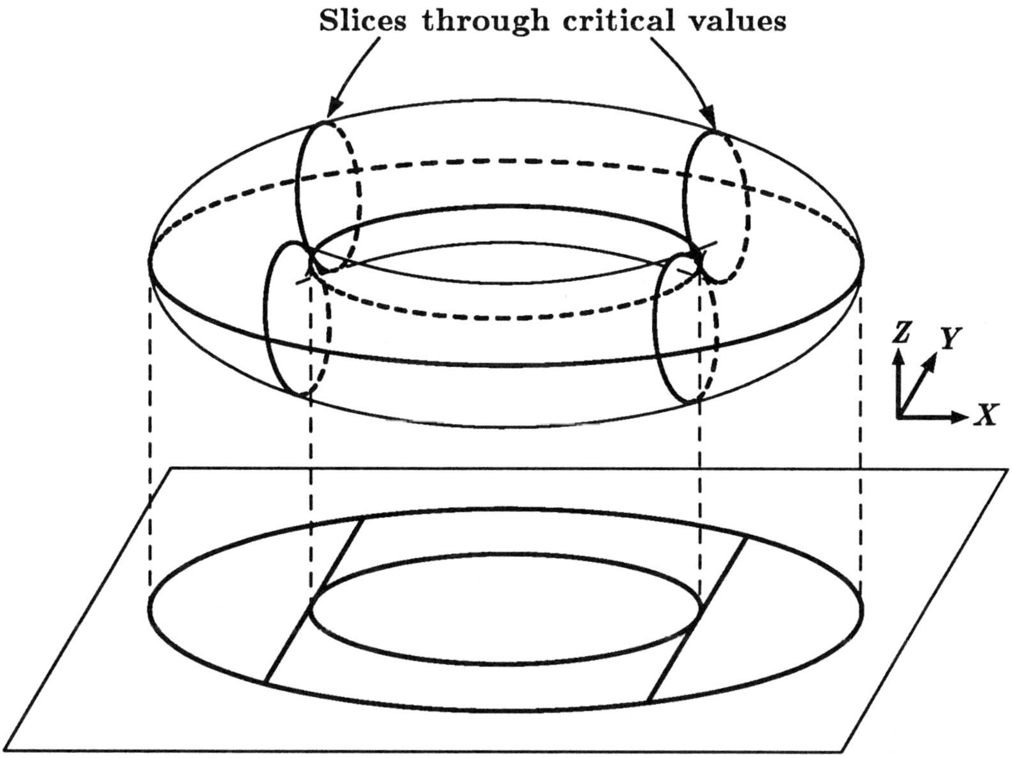

Figure 4.11: The roadmap of the torus, showing recursively computed roadmaps of slices

Then the slice $S|^{v_i}$ is a deformation retract of the "slab" $S|^{[u_{i-1},u_i]}$. This follows from retraction lemma applied twice, i.e. first retract $S|^{[u_{i-1},u_i]}$ onto $S|^{[u_{i-1},v_i]}$, and then retract in the opposite direction onto $S|^{v_i}$. Furthermore, this retraction is compatible with the stratification $\underline{\underline{S}}'$, so that it simultaneously retracts the silhouette. By the hypothesis of this lemma, the retract satisfies the roadmap condition, so if C^i_α is a connected component of $S|^{[u_{i-1},u_i]}$, then $R_0(\underline{\underline{S}}) \cap C^i_\alpha$ is connected.

Now let U be any connected component of S, then U is a union of a finite number of C^i_α, which we denote by $U_1, \ldots U_n$. Now the U_j are closed sets, and can always be ordered so that every U_j for $j > 1$ intersects some U_k with $k < j$. If U_j and U_k intersect, then their intersection is compact, and π_2 will attain an extremum on the intersection, and this point lies in the silhouette. So if U_j and U_k intersect, so do their roadmaps. Thus $R_0(\underline{\underline{S}}) \cap U_1, \ldots R_0(\underline{\underline{S}}) \cap U_n$ is a ordered sequence of sets each of which intersects some earlier set. It follows that their union $R_0(\underline{\underline{S}}) \cap U$, is connected. □

Theorem 4.5.2 *$R_0(\underline{\underline{S}})$ satisfies the roadmap condition.*

Proof. By induction on dimension, the proof follows from the previous lemma if we can show that for $d > 1$, $\underline{\underline{S}}'|^v$ is a Whitney stratification of dimension at most $d - 1$. Now $\underline{\underline{S}}'|^v$ is the intersection of $\underline{\underline{S}}'$ and $\pi_1^{-1}(v)$. Furthermore, if the map π_1 is one-generic, all the critical values v are non-degenerate, so that the critical set consists of isolated points. Now the critical points of $\pi_1|_{\underline{\underline{S}}'}$ all lie on the one dimensional strata in $\underline{\underline{\Sigma}}'$, and since v is a non-degenerate critical value, the intersection of $\pi_1^{-1}(v)$ with these strata is a finite set of points. All other strata meet $\pi_1^{-1}(v)$ transversally, so that their intersection has dimension $d - 1$ and is a Whitney stratification. However, we can trivially extend a Whitney stratification by adding a finite set of points, so by adding the critical points we find that $\underline{\underline{S}}'|^v$ is a Whitney stratification of dimension at most $d - 1$. □

Corollary 4.5.3 *For any stratum $U \in \underline{\underline{S}}$, $\overline{U} \cap R_0(\underline{\underline{S}})$ satisfies the roadmap condition as a subset of $\overline{U}$, where $\overline{U}$ is the closure of U.*

Proof $\overline{U}$ inherits a silhouette-compatible Whitney stratification from $\underline{\underline{S}}'$, and slices are taken at all the critical points of the silhouette of $\overline{U}$, so the above proofs hold. □

In fact, let $R_1(\underline{\underline{S}})$ be a one-dimensional set, then if $\overline{U} \cap R_1(\underline{\underline{S}})$ satisfies the roadmap condition for every $U \in \underline{\underline{S}}$, then $R_1(\underline{\underline{S}})$ satisfies the roadmap condition. The proof follows that of the last lemma. That is, every component of S is a union of a finite number of $\overline{U}_i$, with $U_i \in \underline{\underline{S}}$. These $\overline{U}_i$ can be ordered so that each $\overline{U}_i$ intersects an earlier $\overline{U}_j$. Whenever $\overline{U}_i$ and $\overline{U}_j$ intersect, the intersection contains a stratum of $\underline{\underline{S}}$, which has a connected roadmap, and therefore the roadmaps of $\overline{U}_i$ and $\overline{U}_j$ intersect also. By induction it follows that $R_1(\underline{\underline{S}})$ satisfies the roadmap condition.

This gives us a more local criterion for defining a roadmap, and leads us eventually to a more efficient algorithm.

4.5.2 Linking Curves

We have set ourselves the subgoal of finding a one-dimensional set which satisfies the roadmap condition within the closure of every stratum in $\underline{\underline{S}}$. The first step is to define a link curve for a stratum U which joins any point p in $\overline{U}$ to a distinct point q which is contained in the silhouette of $\overline{U}$. This curve should lie entirely within the slice $\overline{U}|^v$.

Now U is contained in the kernel of a set of n polynomials, with $n =$ codim(U). Similarly, $U|^v$ is contained in $\hat{U}$, which is the intersection of this kernel with the set $\pi_1^{-1}(v)$. In fact $\hat{U} - \{p\}$ is a manifold of codimension $n + 1$, (even if p is a critical point). So the set $\underline{\underline{\hat{U}}} = \{\hat{U} - \{p\}, \{p\}\}$ is a Whitney stratification of $\hat{U}$.

Definition We define $L_0(p, U)$ as the connected component of $R_0(\underline{\underline{\hat{U}}}) \cap \overline{U}$ which contains p, where the silhouette of $\underline{\underline{\hat{U}}}$ is computed using the linear map π_{23}.

Now the link $L_0(p, U)$ contains a point which is extremal wrt π_2. This point, call it p_1, must either lie on the silhouette of U, or it must be at the endpoint of a closed curve segment, which implies that it lies in some stratum U_1, which is adherent to U. vThen the link $L_0(p_1, U_1)$ must contain a point p_2 in the silhouette of U_1, or p_2 must lie in some stratum adherent to U_1 etc. Now we set

$$L_1(p, U) = L_0(p, U) \cup \bigcup_{i<k-1} L_0(p_i, U_i) \tag{4.21}$$

where k is the least i s.t. $p_i \in \Sigma(\overline{U})$. This is well defined because $\dim(U_i) < \dim(U_{i-1})$. Then $L_1(p, U)$ is a connected one-dimensional set which contains both p and a point on the silhouette of $\overline{U}$. This set suffices as the link curve as long as p is a regular point of $\pi_1|_U$. However, if p is a critical point, several components of U may "meet" at p for the first time. That is, let C be the component of $U|^v$ which contains p. $C - \{p\}$ may not be connected, however it is important that we link p to the silhouette of every connected component of $C - \{p\}$. So it is neccessary to link every connected component of $L_0(p, U) - \{p\}$ to the silhouette. Let q_i denote an extremal point wrt π_2 of the ith connected component of $L_0(p, U) - \{p\}$, and let V_i be the stratum which contains q_i. Then we define

$$L_2(p, U) = L_0(p, U) \cup \bigcup_i L_1(q_i, V_i) \tag{4.22}$$

where $L_1(q_i, V_i)$ is empty if q_i is already a silhouette point. One other case where the link curve L_0 is insufficient occurs when the connected component of $U|^v$ which contains p consists of the single point p. This occurs when p is a critical point of $\pi_1|_U$ of index 0 (see [Mil63]). In this case there is no point trying to link p to the silhouette of U, since it consists only of the point p itself. Instead we link p to the silhouettes of all strata to which p is adherent in $\pi_1^{-1}(v)$. In fact it suffices to do this only for the strata of dimension $\dim(U) + 1$ which we denote by W_i. So we set

$$L_3(p, \underline{\underline{S}}) = \bigcup_i L_1(p, W_i) \tag{4.23}$$

and given an arbitrary point p we define:

$$L(p, \underline{\underline{S}}) = \begin{cases} L_1(p, U) & \text{if } p \text{ is a regular point of } \pi_1|_U; \\ L_2(p, U) & \text{if } p \text{ is a critical point of } \pi_1|_U \text{ of index } \neq 0; \\ L_3(p, \underline{\underline{S}}) & \text{otherwise} \end{cases} \tag{4.24}$$

4.5.3 The Full Roadmap

Let $\Sigma = \Sigma(\pi_{12}|_{\underline{\underline{S}}})$ be the silhouette of S, and let $\underline{\underline{S}}'$ be the silhouette-compatible stratification of S. Let P_c denote the set of critical points of $\pi_1|_{\underline{\underline{S}}'}$ *which are not contained in the set of boundary points* B defined earlier. That is

$$P_c = \Sigma(\pi_1|_{\underline{\underline{S}}'}) - \{\partial C | C \in \underline{\underline{\Sigma}}\} \tag{4.25}$$

then the full roadmap $R_1(\underline{\underline{S}})$ is defined by:

$$R_1(\underline{\underline{S}}) = \Sigma \cup \bigcup_{p \in P_c} L(p, \underline{\underline{S}}) \tag{4.26}$$

and we have our main result:

Theorem 4.5.4 *$R_1(\underline{\underline{S}})$ satisfies the roadmap condition.*

Proof Let p_i be the critical points of $\pi_1|_{\underline{\underline{S}}'}$, and let v_i be the corresponding critical values. Assume the v_i are arranged in ascending order. The proof is by induction on the v_i. We assume that $R_1(\underline{\underline{S}})|^{(-\infty, v_{i-1}]}$ satisfies the roadmap condition as a subset of $S^{(-\infty, v_{i-1}]}$, and that the restriction of the roadmap to the closure of a stratum $\overline{U}|^{(-\infty, v_{i-1}]}$ also satisfies the roadmap condition for all $U \in \underline{\underline{S}}$. If we can show that this condition holds for slices $(\infty, v_i]$ we are done. First notice that the result holds for all strata which do not cobound p_i because these strata, along with their silhouettes, may be retracted back to $\underline{\underline{S}}|^{(-\infty, v_{i-1}]}$. For the remaining strata there are four cases:

(i) $p_i \in B = \{\partial C | C \in \underline{\underline{\Sigma}}\}$. Our general position assumption implies that p_i is not also a critical point of $\underline{\underline{\Sigma}}$. Any stratum which cobounds p_i also cobounds some silhouette curve which cobounds p_i. Since $d\pi_1(T_x C)$ is non-zero for any such curve C, there must be nearby points in $\pi_1^{-1}(v_i - \epsilon)$ on all curves for sufficiently small ϵ. It follows that all these curves link p_i to the slice $\underline{\underline{S}}|^{(-\infty, v_{i-1}]}$.

(ii) $p_i \in U$ is a regular point of $\pi_1|_U$ (notice that this includes cusp points). Since p_i is linked to a silhouette point of U or of a stratum in the boundary of U, the roadmaps of all strata which cobound p_i remain connected, because these strata must also have the silhouette point in their boundary (this follows by transitivity of the adherence relation, since the stratification $\underline{\underline{S}}$ satisifies the frontier condition).

(iii) $p_i \in U$ is a critical point of $\pi_1|_U$ of index $\neq 0$. In this case, several pieces of U may "come together" at p_i. i.e. there may be several connected components of $U|^{(-\infty, v_i]} - \{p_i\}$ which cobound p_i. However, every component must contain a component of $L_0(p, U)$ linking it to a point q_i and thence to a silhouette point in that component. Thus the roadmap of $\overline{U}|^{(-\infty, v_i]}$ satisfies the roadmap condition, and the same holds for all strata which cobound U, again by transitivity of the adherence relation.

(iv) $p_i \in U$ is a critical point of $\pi_1|_U$ of index 0. Here the problem is not with the stratum U itself, since locally U consists only of the point p_i which equals its own roadmap, but with the strata W_i which cobound U, since p_i is an isolated roadmap point in the boundary of these strata. It is necessary to connect p_i to distinct silhouette points on each W_i, but $L_3(p, \underline{\underline{S}})$ does this. Thus the roadmaps of all W_i satisfy the roadmap condition, and the same holds for strata which cobound the W_i, once again by transitivity of adherence. □

4.6 The Roadmap Algorithm

We define the algorithm bottom-up, describing first the algebraic algorithms for computing equations for curves and points, and later the geometric algorithms for construction of the adjacency lattice.

Algorithm 4.1
Input: *Polynomials* $f_1, \ldots f_n$ *in* r *variables* $x_1, \ldots x_r$ *of degree* d *defining a manifold* M, *and a linear map* $\pi_{12} : \Re^r \rightarrow \Re^2$.
Output: *a single polynomial* $h(x_1, x_2)$ *s.t.* $\ker(h) \supset \pi_{12}(\Sigma(\pi_{12}|_M))$

Description

We use the slice lemma ([slice-lemma]) to infer that the critical set of $\pi_{12}|_M$ is the union of the sets of critical points of $\pi_2|_{M_c}$ for all slices $M_c = M \cap \pi_1^{-1}(c)$.

We can now apply the tube theorem (4.4.12) directly and we define a polynomial

$$g = \sum_{i=1,\ldots,n} f_i^2 \qquad (4.27)$$

If we define $K_c = g^{-1}(\epsilon) \cap \pi_1^{-1}(c)$ then in any slice, for all sufficiently small ϵ, there will be a critical point of $\pi_2|_{K_c}$ near each non-degenerate critical point of $\pi_2|_{M_c}$. The union of the critical points of $\pi_2|_{K_c}$ for all slices is just the critical set of $\pi_{12}|_{g^{-1}(\epsilon)}$. We need to choose ϵ small enough to succeed in all of the slices, but we can make use of the fact that the critical set of $\pi_{12}|_M$ is compact (or at least the part we are interested in is a compact subset). This shows that the tube theorem extends to maps into $\Re^2$, and for sufficiently small ϵ, there is a critical point of $\pi_{12}|_{g^{-1}(\epsilon)}$ arbitrarily close to every non-degenerate critical point of $\pi_{12}|_M$.

From here we can proceed in one of two ways. We could try to directly eliminate $x_3, \ldots, x_r$ from the system of equations

$$g = \epsilon \quad \frac{\partial g}{\partial x_3} = 0, \;\cdots\; \frac{\partial g}{\partial x_r} = 0 \tag{4.28}$$

since these equations define the critical set of $\pi_{12}|_{g^{-1}(\epsilon)}$. But we need to be cautious in using the multivariate resultant. The resultant gives a condition for solvability over the complex numbers, while we are interested only in the real solution set. However, it turns out that there is a version of Sard's theorem that applies to algebraic sets over the complex numbers, [Mum] lemma 3.7. This states that with the same definition of critical points and values as in (4.28), the set of regular values of a map is dense in its range. So the projection of the complex critical set is a measure zero subset of $\Re^2$. Since the complex critical set contains the real critical set, their projections both have dimension one. At worst the projection of the complex solution set will contain some additional curves, which can be detected and removed later.

A more serious problem arises when we homogenize the system (4.28). For this system, homogenization introduces extra solutions at infinity that have high dimension, and cause the resultant to vanish identically. The correct projection of the critical set can be salvaged using the recent result in [Ca88b] on generalized characteristic polynomials. This will give the projection as the zeros set of a polynomial $R(x_1, x_2, \epsilon)$, and we are interested in the shape of the curve $R = 0$ in x_1-x_2 space as ϵ tends to zero. Arranging R in powers of ϵ, the lowest degree term in ϵ determines the shape of the curve in the limit. So the coefficient of this lowest degree term, which will be a polynomial in x_1 and x_2, is the polynomial h which we are looking for.

A more elementary method that does not use generalized characteristic polynomials involves changing the definition of g slightly, so that the excess solutions at infinity are removed. We define

$$g' = \frac{\sum f_i^2}{1 + \sum x_j^{2d}} \tag{4.29}$$

and then in a small neighborhood of any point $p \in M$, g' closely approximates a constant times g. The steps in the proof of the tube theorem all go through for g', and so we may find critical points by solving the system

$$g' = \epsilon \quad \frac{\partial g'}{\partial x_3} = 0, \cdots \frac{\partial g'}{\partial x_r} = 0 \tag{4.30}$$

and we claim that after homogenization, the solution set of the system (4.30) is one-dimensional. To prove this, we look at the solution set in various affine subsets of projective r-space. The ith affine subset has $x_i \neq 0$, and together they cover projective r-space. Almost every ϵ is a regular value of g' within all of the slices, and for these values, the preimage is a manifold N in projective space. Applying the algebraic version of Sard's theorem to the restriction of the projection map $a|_N$ (within each affine subset) shows that the regular values are a dense subset. This implies that the resultant does not vanish identically. It will be a polynomial $R(x_1, x_2, \epsilon)$, and as before, $h(x_1, x_2)$ is the coefficient of the lowest degree term in ϵ of $R(x_1, x_2, \epsilon)$.

Computation of $h(x_1, x_2)$ requires the computation of the resultant of a system of $r - 2$ polynomials of degree $2d - 1$ and one of degree $2d$. Because of the introduced ϵ, the constant terms in the recurrence (3.11) are non-vanishing (in fact most are monic polynomials in ϵ), so we can use method 1 for resultant computation. Each matrix in the computation has size at most $O((6d)^r)$ and we need compute only the highest $(2d)^r$ terms of the matrix coefficients as polynomials in ϵ. This can be done in $O(r^2(12d^2)^{3r} \log dw)$ binary operations.

Algorithm 4.2
Input: *A linear map* $a : \Re^r \rightarrow \Re^2$ *and polynomials* $f_1, \ldots, f_n$ *and* f_k *in* $x_1, \ldots, x_r$ *of degree* d. *The zero set of* $f_1, \ldots, f_n$ *is assumed to be a manifold* M.
Output: *The intersection points of the curve* $C = \Sigma(a|_M)$ *and the hypersurface* $f_k = 0$.

Description This algorithm is a hybrid of algorithm 3.2 and algorithm 4.2. First we form a polynomial g from $f_1, \ldots, f_n$ using (4.4.12). The solution set of the system (4.28) is a smooth one-dimensional curve, and the intersection of this curve with the hypersurface $f_k = 0$ generically consists of a finite number of points, and as $\epsilon \rightarrow 0$ these points converge to the solution points we seek. We find these points using the u-resultant. We introduce a linear form in three variables as in (3.41), and find the resultant of the polynomial system (4.28) and $f_k = 0$ and one of the forms (3.41). This is a system of $r+1$ equations in $r+1$ unknowns (once the homogenizing variable is added). We arrange the resultant in powers of ϵ, and take the coefficient of the lowest power.

Once again, the intermediate polynomials in the resultant computation are non-vanishing, so we can use method 1 to compute them. This step takes $O(r^3(12d^2)^{3r} \log dw)$ binary operations. We use the substitution method described in algorithm 3.2 to find solution vectors, and this phase dominates the running time of the algorithm. We can apply the analysis of algorithm 3.2 to algorithm 4.2, but where the degrees of the polynomials are doubled. This gives an $O(r^3(2d)^{5r} \log^2 dw)$ bound for the number of binary operations.

Algorithm 4.3
Input: *A linear map* $a : \Re^r \rightarrow \Re^2$, *and polynomials* $f_1, \ldots f_n$, *in* r *variables* $x_1, \ldots x_r$ *of degree* d *defining a semi-algebraic set* S, *such that the first* $m < r$ *polynomials define a manifold* M.
Output: *The adjacency graph of the smooth curve* $C = \Sigma(a|_M)$ *and all the intersection points of this curve with the hypersurfaces* $f_i = 0$ *for* $i = m+1, \ldots, n$. *Each point and curve segment is labelled with the signs of all the* f_i.

Description We first call algorithm 4.1 to compute the projection of the curve C. Then we use a variation of a line sweep method to order intersection points along each connected component of C. The algorithm is essentially the same as that described in [Ca85]. Algorithm 4.1 returns a polynomial h in two variables say x_1 and x_2. We first ensure that h is square-free. Then we compute the Sturm sequence of h as a polynomial in x_1. The last term in this sequence is a polynomial $D(x_2)$ in x_2 only, called the discriminant of h. If h is square-free, then the discriminant is a polynomial in x_2 having only a finite number of zeros. For x_2 in between these zeros, the curve $h = 0$ has a fixed number of solutions in x_1, and these form smooth curves as x_2 is varied. They may therefore be uniquely ordered by x_1 coordinate in this range of x_2.

Given the x_1, x_2-coordinates of an intersection point p found using algorithm 4.2, we can determine from the signs of the Sturm sequence at p whether the projection of the point lies on a curve, and if it does, the number of the curve in the x_1-ordering. We can check by substitution into the original system of polynomials whether the point actually lies on the curve. Thus we can completely construct the adjacency graph of points and curve segments in between zeros of $D(x_2)$. At the zeros of $D(x_2)$ we can make use of the fact that the projection of a smooth curve into the plane is generically an immersion with normal crossings. This means that there are only two types of event that can occur at a zero α of x_2: (i) two curve segments meet smoothly, in which case there will be either be two more or two less zeros of $h(x_1)(x_2)$ for $x_2 > \alpha$ than for $x_2 < \alpha$. (ii) two curve segments cross, in which case there will be the same number of curves on either side of α. Both these cases are easily recognised, and the Sturm sequence indicates which curve segments join or cross. Thus we can unambiguously identify curve segments across zeros of $D(x_2)$, and therefore compute the complete adjacency graph of the curve and all intersection points.

Since the graph has pure degree two (C is a smooth curve), there is a linear ordering of points along each component of the graph. Exactly one polynomial changes sign at each point, so by using a persistent data structure, (e.g. an AVL-tree without mutation), we can store the signs of all n polynomials at each point at $O(\log n)$ cost per point, and retrieve them in $O(\log n)$ time.

We need $n - m$ calls to algorithm 4.2, a cost of $O(nr^3(2d)^{5r} \log^2 dw)$. We use approximate arithmetic once again to evaluate the Sturm sequences, and in fact the gap required, $O(r(2d)^{3r} \log dw)$ is the same as that used in algorithm 4.2, and so the values supplied by that algorithm are accurate enough. Evaluation of the entire sequence at a point takes time $O(r^2(2d)^{4r} \log^2 dw)$, and there are less than $n(2d)^r$ points. The Sturm evaluation and point ordering phase therefore takes $O((n \log n)r^3(2d)^{5r} \log^3 dw)$ binary operations, including time to generate the query structure. This dominates the running time of the algorithm.

Algorithm 4.4 *(Basic Roadmap)*
Input: *Polynomials $f_1, \ldots, f_n$ in r variables of degree d defining a semi-algebraic set S. The f_i are assumed to be in general position, so that the stratification $\underline{\underline{S}}$ of S into sign-invariant sets is a Whitney stratification.*
Output: *The adjacency graph and point coordinates of the basic roadmap $R_0(\underline{\underline{S}})$.*

Description. We first choose a one-generic projection map a. Then the basic roadmap consists of the silhouette $\Sigma(a|_{\underline{\underline{S}}})$ and the basic roadmap of slices through critical points on the silhouette. To compute the silhouette, we call algorithm 4.3 on all subsets of $m < r$ polynomials. The adjacency graph of the silhouette is readily constructed from the graphs of individual curves in two steps: (i) identifying corresponding points on curves on strata of the same dimension, and (ii) find the position of points on curves of lower dimensional strata. Step (i) can be accomplished in time linear in r if points are stored lexicographically ordered by coordinates, and in arrays indexed by the polynomials that are zero at the point. For step (ii) we simply locate the point on the curve as though it were an intersection point for that curve. Overall, the adjacency graph of the silhouette can be constructed in $O(\binom{n}{r-1}(n \log n) r^4 (2d)^{5r} \log^3 dw)$ binary operations.

Now for the recursive step, we need to count the number of critical points on silhouette curves. First of all there are $\binom{n}{r-1}$ curves and each will have at most $(2d)^{2r}$ critical points. This follows from the maximum degree of the curves. So if $T_{R_0}(k)$ is the number of binary operations needed in the basic roadmap in k dimensions, then we have the following recurrence:

$$T_{R_0}(k) = \binom{n}{k-1}(2d)^{2k} T_{R_0}(k-1) \; + \; O(n^k \log n \frac{k^4}{k!}(2d)^{5k} \log^3 dw) \quad (4.31)$$

and solving, we find that the recursive step dominates, and that the basic roadmap in r dimensions requires

$$T_{R_0} = O(n^{\frac{1}{2}r(r-1)}(2d)^{r(r+1)}) \quad (4.32)$$

binary operations.

Algorithm 4.5 *(Point Linking)*
Input: *The coordinates of a point* $p \in \Re^r$*, polynomials* $f_1, \ldots, f_n$ *of degree* d *in* r *variables, a linear map* a *and the silhouette (or partial roadmap) of* $\underline{\underline{S}}$ *(including adjacency graph).*
Output: *The link curve* $L(p, \underline{\underline{S}})$ *defined in section 5.2, and a modified adjacency graph which includes this curve.*

Description. We must first determine whether p is a critical point. We test the signs of all the f_i at p, and those which are zero (say m of them) define the local geometry of the stratum U containing p. If the projection of the tangent space of this stratum is not surjective (checked using equation (4.9), then p is a critical point. To find the index of p, we examine the local direction of the flow (section 3) induced by a vector field in the a_1 direction. The flow at each point is a vector whose inner product with a vector in the a_1 direction is always non-negative, and which is zero only at critical points. By solving a simple $m \times m$ system of linear equations, we can explicitly define (giving the components as polynomials) a vector field parallel to the flow which also vanish only at critical points, and has non-negative inner product with a_1. This vector field has degree $3dm$. Taking partial derivatives of the vector field with respect to the x_i, we obtain an $r \times r$ matrix H, which describes the local change in the flow field near p. We can transform this matrix to an $(r - m) \times (r - m)$ matrix H' using a local basis for the tangent space to U at p. Then p is a critical point of index 0 iff all the eigenvalues of H' are positive.

By the above tests we can determine which type of linking curve is needed at p. We analyse here only the most complicated case, which is where p is a critical point of index 0. Thus we must link p to the silhouettes of all strata adherent to p of dimension one greater than U. First we call algorithm 4.4 on all subsets of size $m-1$ of the m polynomials which define U. Then to each adjacency graph returned by one of these calls, we add the intersections with all constraints $f_k = 0$. However, rather than maintaining a sorted structure as in algorithm 4.3, we keep track only of those intersection points which are closest to p in the linear ordering along some curve through p (in fact we need only keep track of the closest point on one such curve). If there is a direct path from p to a silhouette point (i.e. no intermediate intersection point with some $f_k = 0$) then we are done for that graph, and we add it to the main graph. Otherwise, we must compute the roadmap of a lower-dimensional stratum. We repeat this at most $r - m$ times until we obtain a silhouette point.

Thus there are $m(r - m)$ calls to the basic roadmap algorithm in dimension $r - 1$, each with at most m polynomials. From algorithm 4.4, these calls have a total cost of $O(r^2 r^{\frac{1}{2}(r-1)(r-2)}(2d)^{r(r-1)})$. The other step is addition of new nearest intersection points with some $f_k = 0$, which from the description of algorithm 4.3 takes time $O(r^2(2d)^{4r}\log^2 dw)$ operations per point. Counting the number of possible intersections using the recurrence (4.31) we find that there will be at most $O(nr^2 r^{\frac{1}{2}(r-1)(r-2)}(2d)^{r(r-1)})$ such intersection points. This phase dominates the running time, so that linking critical points of index 0 takes time:

$$T_L = O(nr^4 r^{\frac{1}{2}r(r-1)}(2d)^{r(r+3)}\log^2 dw) \tag{4.33}$$

binary operations. All other types of link require less time than this.

Algorithm 4.6 *(Full Roadmap)*
Input: *A Formula defining a semi-algebraic set S, with defining polynomials in general position.*
Output: *The full roadmap $R_1(\underline{\underline{S}})$ of the stratification $\underline{\underline{S}}$ of S into sign-invariant sets.*

Description
The algorithm has two phases. First we choose a generic linear map a and compute the silhouette Σ of $\underline{\underline{S}}$. From the description of algorithm 4.4, this requires $O(n^r \log n \frac{r^4}{r!}(2d)^{5r}\log^3 dw)$ binary operations.

Then we call algorithm 4.5 (linking) on all critical points of $a_1|_\Sigma$. Now there are $\binom{n}{r-1}$ silhouette curves, and each may have at most $(2d)^{2r}$ critical points. So the total time for linking is $O((n^r)\frac{r^4}{r!}r^{\frac{1}{2}r(r-1)}(2d)^{r(r+5)}\log^2 dw)$ binary operations. So overall the full roadmap algorithm requires

$$T_{R_1} = O\left((n^r)\frac{r^4}{r!}(\log n(2d)^{5r}\log^3 dw + r^{\frac{1}{2}r(r-1)}(2d)^{r(r+5)}\log^2 dw)\right) \tag{4.34}$$

binary operations.

Chapter 5

Performance Improvements

In this chapter[1] we study heuristic improvements for motion planning algorithms. We consider only provably good heuristics, which should improve average case running time without sacrificing accuracy or completeness. We define a certain kind of Voronoi diagram, which is an $r-1$ dimensional subset of the r dimensional set of free configurations. The main result of section 5.1 is that this diagram is complete for motion planning. The Voronoi diagram construction has several advantages. Firstly, when used with the roadmap algorithm, it gives paths with maximal clearance from obstacles, whereas the roadmap of free space contains curves that follow the boundary of free space, i.e. the roadmap algorithm may produce paths that touch obstacles. Secondly, unlike free space, the diagram has no interior. The roadmap algorithm, at least in its basic form, would generate numerous curves on the interior of free space that are unnecessary for preserving connectivity within each connected component of free space. Finally, there is empirical evidence that the Voronoi diagram of some obstacles in r dimensions has lower complexity than a general arrangement of the constraints that bound the obstacles.

[1] Parts of this chapter describe joint work with Bruce Donald of Cornell University

5.1 The Simplified Voronoi Diagram

The Voronoi diagram has proved to be a useful tool in a variety of contexts in computational geometry. Our interest here is in using the diagram to simplify the planning of collision-free paths for a robot among obstacles, the generalized movers' problem. The Voronoi diagram, as usually defined, is a *strong deformation retract* of free space so that free space can be continuously deformed onto the diagram. This means that the diagram is complete for path planning, i.e. Searching the original space for paths can be reduced to a search on the diagram. The diagram has dimension one less than the original space. Reducing the dimension of the set to be searched usually reduces the time complexity of the search. Secondly, the diagram leads to robust paths, i.e. paths that are maximally clear of obstacles.

The Voronoi diagram generated by a set of points in a Euclidean space partitions the space into convex regions which have a single nearest point under some (usually L_2) metric. A generalized Voronoi diagram can be defined for points and line segments in the plane [LD] which partitions the plane into (generally non-convex) regions. In both cases the diagram is defined to be the set of points equidistant from two or more object features under the appropriate metric. This construction has proved to be useful for motion planning among a set of obstacles in configuration space (see [OY], [OSY], [Yap] and the textbook of Schwartz and Yap [SY] for an introduction and review of the use of Voronoi diagrams in motion planning). Its virtue for motion planning is that the diagram is a strong deformation retract of free space, i.e. the space outside the obstacles can be continuously deformed onto the diagram. To find a path between two points in free space, it suffices to find a path for each point onto the diagram, and to join these points with a path that lies wholly on the diagram.

The simplified diagram has lower algebraic complexity than the L_2 diagram. For example, in $\Re^3$, the L_2 diagram for polyhedral obstacles consists of quadric sheets; the simplified diagram is piecewise linear. In $\Re^2$, the simplified diagram for polygonal obstacles is an arrangement of straight line segments, see figure 5.2. In general, the simplified diagram has the same degree as the obstacle constraint equations. However, it may not have linear size in the worst case.

The simplified Voronoi diagram is also defined for the six-dimensional

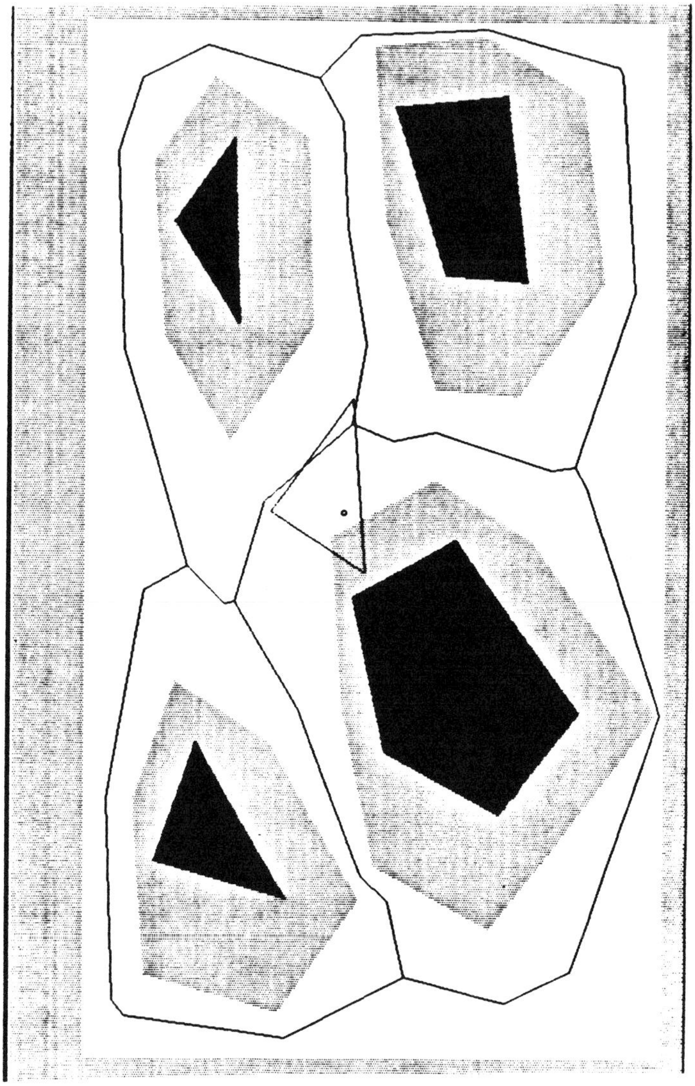

Figure 5.1: The simplified Voronoi diagram in the plane. The central triangle is the object, the circle is its reference point. Physical obstacles are shown in black, and the configuration space obstacles are grey.

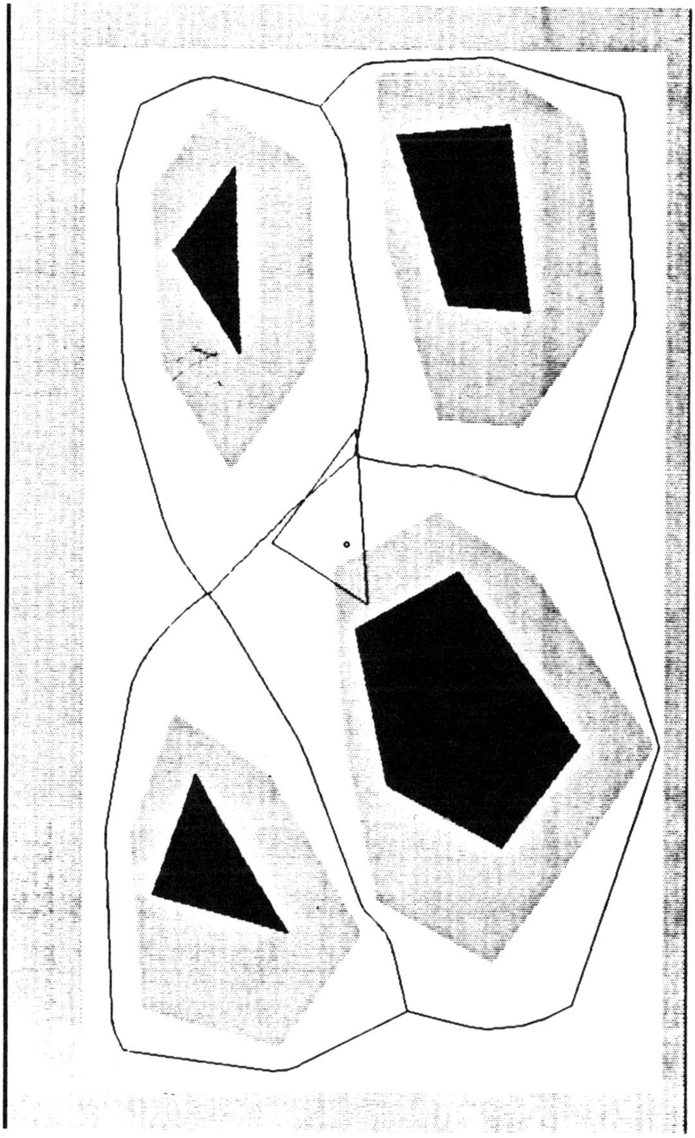

Figure 5.2: True Voronoi diagram for the same obstacles.

configuration space of an arbitrary 3D polyhedron moving amidst 3D polyhedral obstacles. Our definition elaborates a suggestion of Donald [Don] and Canny [Ca85], who describe certain Voronoi-like properties of the algebraic set $\bigcup_{i \neq i'} \ker(C_i - C_{i'})$ for a set of algebraic constraints $\{ C_i \}$. In this chapter, we consider the configuration space of a polyhedral object with translational and rotational degrees of freedom. The simplified Voronoi diagram has the same algebraic complexity as the resulting configuration space obstacle boundaries.

The completeness property holds for the simplified diagram when the polynomial equations that define the configuration obstacle boundary have unit gradients throughout. The diagram has the same degree as these normalized constraints. (The *degree* of the diagram is the degree of the defining equations). Thus in $\Re^2$ and $\Re^3$ the simplified diagram has degree one whereas the Euclidean diagram has degree two. However, note that the Euclidean diagram in $\Re^3$ has curves of degree four and vertices of degree eight, whereas the simplified diagram has linear features only. In the configuration space $\Re^2 \times S^1$ of a planar polygon, the simplified diagram has degree three, whereas the Euclidean diagram has degree six. In the six-dimensional configuration space $\Re^3 \times SO(3)$ of a 3D polyhedron, both the simplified and Euclidean diagram have degree ten (It is higher than three because of the presence of type-C constraints, which require the use of normalizing polynomials). The simplified Voronoi diagram has a relatively small stratification, $O(n^7)$ strata, rather than $O(n^{12})$, which is the naive bound. The worst-case computing time for the roadmap of the diagram is $O(n^7 \log n)$, but we believe that in practice it will be lower than this. We conjecture that the diagram in fact has no more than $O(n^6)$ strata.

The completeness proof in this chapter applies to the configuration space constraints for a polyhedron in free space, whose configuration space is $\Re^3 \times SO(3)$. The results can readily be specialized to the case of a polygon moving in two dimensions with rotations, whose configuration space is $\Re^2 \times S^1$. The proof assumes that these constraints are normalized (the gradient of each constraint has magnitude one everywhere), although the simplified diagram can also be defined for unnormalized configuration space constraints. In the lower-dimensional configuration spaces, the constraints may be normalized with no increase in algebraic complexity. In $\Re^3 \times SO(3)$, a quaternion representation of rotation gives constraints of degree three in the configuration space coordinates. In fact the constraints are simultaneously quadric in the quaternion coordinates and linear in the position coordinates.

Some of the constraints (called type C, see chapter 2) do not a priori satisfy the normalization condition, but they can be normalized by dividing by a polynomial factor. Since this increases the degree of the equations defining the diagram (to an effective degree of ten, the same as the Euclidean diagram), we suggest instead that they be left unnormalized. We believe that the cases where this causes the diagram to become disconnected are rare, because at reasonable distances the normalized type A constraints determine the value of the distance function. We think that leaving the type C constraints unnormalized is a good compromise between efficiency and provable completeness.

5.1.1 A Generalized Distance Function

For motion planning, the configuration of object A is variable. For a polyhedron in three dimensional space, the configuration contains a position $\mathbf{x} \in \Re^3$ and an orientation $\mathbf{q} \in SO(3)$ component, where $SO(3)$ is the group of three-dimensional rotations. Thus the face normals, vertex locations, and edge directions of A are all functions of $\mathbf{x}$ and $\mathbf{q}$. The predicate (2.10) from chapter 2 is also a function of the configuration $(\mathbf{x}, \mathbf{q})$, and is of the form

$$F(\mathbf{x}, \mathbf{q}) = \bigwedge_i \bigvee_j \bigwedge_k \bigvee_l (C_{ijkl}(\mathbf{x}, \mathbf{q}) > 0) \tag{5.1}$$

Where the $C_{ijkl}(\mathbf{x}, \mathbf{q})$ come from type A or type C predicates as defined in Chapter 2. They are algebraic with the quaternion representation of rotation. The set of overlap configurations is called the *configuration obstacle* and is denoted $CO = \{(\mathbf{x}, \mathbf{q}) | \neg F(\mathbf{x}, \mathbf{q})\}$. It may be thought of as a physical obstacle in configuration space to be avoided by a path planner. We now observe that by letting positive real values represent logical one, and non-positive values represent logical zero, that the min function implements logical AND, and the max function implements logical OR. Thus an equivalent form to (5.1) is

$$F(\mathbf{x},\mathbf{q}) = \left(\left(\min_i(\max_j(\min_k(\max_l C_{ijkl}(\mathbf{x},\mathbf{q})))) \right) > 0 \right) \tag{5.2}$$

which suggests that the quantity

$$\rho(\mathbf{x},\mathbf{q}) = \min_i(\max_j(\min_k(\max_l C_{ijkl}(\mathbf{x},\mathbf{q})))) \tag{5.3}$$

can be used as a measure of distance from the configuration obstacle, because it varies continuously through configuration space, is positive at configurations outside CO, and non-positive at configurations inside CO. Thus the configuration obstacle can be rewritten as $CO = \{\mathbf{p} \mid \rho(\mathbf{p}) \leq 0\}$, and its complement, the set of points in free space can be written $F = \{\mathbf{p} \mid \rho(\mathbf{p}) > 0\}$.

In order to define the Voronoi diagram under the distance measure ρ, we need a notion of closest feature. The closest features to a configuration $(\mathbf{x},\mathbf{q})$ are those C_{ijkl} which are *critical* in determining the value of $\rho(\mathbf{x},\mathbf{q})$, that is, small changes in the value of C_{ijkl} cause indentical changes in the value of ρ.

Definition.

A constraint $C_{i_0 j_0 k_0 l_0} \in \{C_{ijkl}\}$ is *critical* at a configuration $(\mathbf{x},\mathbf{q})$ if the value of $C_{i_0 j_0 k_0 l_0}(\mathbf{x},\mathbf{q})$ equals the maximum (or minimum) value of every max (resp. min) ancestor of $C_{i_0 j_0 k_0 l_0}$ in the min-max tree in (5.3). i.e.

$$C_{i_0 j_0 k_0 l_0}(\mathbf{x},\mathbf{q}) = \max_l C_{i_0 j_0 k_0 l}(\mathbf{x},\mathbf{q}) = \min_k(\max_l C_{i_0 j_0 kl}(\mathbf{x},\mathbf{q})) = \ldots \tag{5.4}$$

Now we have

Definition.

The *simplified Voronoi diagram* V is the set of configurations in free space F at which at least two distinct constraints are critical.

It should be clear from the definition of criticality that V is semi-algebraic if the constraints C_{ijkl} are algebraic. V is closed as a subset of free space, although it is not closed in configuration space. Notice that V has no interior, since it is contained inside a finite set of bisectors, each of which has no interior. A bisector is the zero set of $(C_{i_0 j_0 k_0 l_0} - C_{i_1 j_1 k_1 l_1})$ for some pair of distinct constraints. It will prove useful to subdivide the Voronoi diagram into two parts:

Definition.

The *concave part* of the Voronoi diagram V denoted conc(V) is the set of configurations in F where two distinct constraints are critical, and the lowest common ancestor of these constraints in the min-max tree of (5.3) is a min-node.

The *convex part* of the Voronoi diagram V denoted conv(V) is the set of configurations in F where two distinct constraints are critical, and the lowest common ancestor of these constraints is a max-node.

Notice that these two definitions are not mutually exclusive, because there may be points where more than two constraints are critical, and which satisfy both definitions. Thus conv(V) and conc(V) may overlap.

5.1.2 Completeness for Motion Planning

Our key result is that any path in F with endpoints in V can be deformed (in F) to a path with the same endpoints lying entirely in V. We start with a path $p : I \rightarrow \Re^3 \times SO(3)$ lying in free space, $p(I) \subset F$, where $I = [0, 1]$ is the unit interval.

First we assume *wlog* that p intersects V at a finite number of points. We can do this because, as defined in the previous section, V is a semi-algebraic set, and by Whitney's [Wh57] result, it can be split into a finite number of manifolds, or *strata*. Since V has no interior, all these manifolds have codimension at least one. For *any* path p there is a path p' arbitrarily close to p, which is an embedding of I, and so $p'(I)$ is a 1-manifold. Almost every perturbation of p' intersects all of the strata transversally, and therefore at a finite number of points. We can choose such a perturbation to be arbitrarily small, in particular, smaller than the minimum distance from $p'(I)$ to CO. Such a perturbation gives a new path p'' which is path homotopic to p, and which has finite intersection with V.

So we assume that p has m intersections with V, and that these occur at points $p(x_i)$ with $x_1, \ldots x_m \in I$, and $x_1 = 0$ and $x_m = 1$. We then break each interval $[x_i, x_{i+1}]$ in half, giving us two intervals sharing an endpoint. Thus we now have $2m-2$ intervals each of which intersects V at only one endpoint. Below we give homotopies for each of these intervals which continuously deform the image of the interval onto V. Since all these homotopies agree at their endpoints, they can be pasted together to give us a global homotopy which deforms p onto V. For simplicity, we assume that each path segment is parametrized in the range $I = [0, 1]$ and that $p(0) \in V$.

The motion constraints C_{ijkl} are either A or C predicates, (2.3) and (2.7), and all can be written in the general form below, called the parametrized plane equations by Lozano-Perez [Loz]:

$$C_{ijkl}(\mathbf{x}, \mathbf{q}) = \mathbf{N}_{ijkl}(\mathbf{q}) \cdot \mathbf{x} + c_{ijkl}(\mathbf{q}) \tag{5.5}$$

Where $\mathbf{N}_{ijkl}(\mathbf{q}) \in \Re^3$ and $c_{ijkl}(\mathbf{q}) \in \Re$ are both continuous functions of $\mathbf{q}$. We assume that the C_{ijkl} are normalized so that $|\mathbf{N}_{ijkl}(\mathbf{q})| = 1$ for all $\mathbf{q}$. Our objective is to continuously deform the path p onto the diagram, and we use the $\mathbf{N}_{ijkl}$ as 'normals' to push a point on the path $p(I)$ away from the critical C_{ijkl}. We assume that the set of positions is bounded by some set of constraint 'walls' of the same form as (5.5), so that a point can be displaced only a finite distance in free space. We also assume that the workspace has unit diameter.

General position assumptions. The construction requires the following general position assumptions. First, suppose C_{ijkl} is type C predicate (2.7). Then $\mathbf{N}_{ijkl}(\mathbf{q}) = d_A(\mathbf{q}) \times d_B$. To normalize $\mathbf{N}_{ijkl}$, we must divide by its magnitude, which must remain non-zero. Hence the set of configurations

$$\{\, \mathbf{q} \mid d_A(\mathbf{q}) \times d_B = 0 \,\} \subset SO(3) \tag{5.6}$$

must not intersect the image of p. However, (5.6) is clearly of codimension 2 in $SO(3)$, and hence by Sard's lemma there is always an arbitrarily small perturbation of p which avoids (5.6).

The second general position assumption relates to configurations where the bisector of two constraints is not well-defined. This occurs on sets of the form

$$\{ (\mathbf{x}, \mathbf{q}) \mid (\mathbf{N}_{ijkl}(\mathbf{q}) = \pm \mathbf{N}_{i'j'k'l'}(\mathbf{q})) \} \tag{5.7}$$

which are also of codimension 2 in $\Re^3 \times SO(3)$. We give a proof for the "+" case, and the $-$ case follows with minor changes.

Consider a map from $\Re^3 \times SO(3)$ to the product $S^2 \times S^2$ which represents the values of the two normals. The diagonal of this space is the set where the two normals agree, and has codimension 2. If the maps are generic (specifically, if they are transversal to the diagonal set), then the preimage of the diagonal set (which is the "bad" set) will also have codimension 2.

We must define a homotopy that deforms a path segment $p(I)$ onto the diagram. Each path segment is assumed to have its endpoint $p(0)$ in V. We need two different homotopies, depending on whether $p(0) \in \mathrm{conc}(V)$, which is the simplest case, or $p(0) \in \mathrm{conv}(V)$.

Notice that since $p(0)$ is the only point on the path which is in V, there is exactly one constraint which is critical at all configurations in $p(0,1]$ (since constraints change value continuously along the path and for another constraint to become critical, it must first equal the first constraint, which can only occur at points on the diagram). Let the critical constraint be C_{ijkl}. We define a homotopy $J_0 : I \times I \to \Re^3 \times SO(3)$ as

$$J_0(t, u) = p(t) + u\mathbf{N}_{ijkl}(\pi_q(p(t))) \tag{5.8}$$

where $\pi_q(p(t))$ is the orientation component of $p(t)$ and the addition symbol means we add the vector quantity $u\mathbf{N}_{ijkl}(\pi_q(p(t))$ to the position component of $p(t)$. The deformation above pushes points beyond the diagram, so we define a second homotopy $J_1 : I \times I \to F$.

$$J_1(t, u) = J_0(t, \min(u, u_c(t))) \tag{5.9}$$

$$\text{where} \quad u_c(t) = \inf\{ s \mid J_0(t, s) \in \mathrm{conc}(V) \}$$

Note that u_c is bounded above by 1 assuming the workspace has unit diameter (obviously, we can scale the value of u if the diameter is larger than one). Recall from the definition of homotopy, that J_1 is a homotopy of p

and a path p' if J_1 is continuous and $J_1(t, 0) = p(t)$ and $J_1(t, 1) = p'(t)$. The homotopy J_1 suffices to map paths with one endpoint in conc(V) onto V:

Lemma 5.1.1 *Let $p : I \to \Re^3 \times SO(3)$ be a path having $p(0) \in V$ and no other points in V. Then J_1 is a homotopy of p and a path p' such that $p'(I) \subset \text{conc}(V)$. Furthermore $p'(0) = p(0)$ if $p(0) \in \text{conc}(V)$.*

Proof. From the definition of J_1 we have $J_1(t, 0) = p(t)$ and $J_1(t, 1) \in$ conc(V). Also if $p(0) \in \text{conc}(V)$ then $u_c(0) = 0$ so $p'(0) = p(0)$. It remains to show that J_1 is continuous. First we notice that J_0 is continuous. Continuity of J_1 follows if we can show that $u_c(t)$ is continuous. Now $J_0(t, u_c(t))$ is contained in the zero sets of all bisectors $\{(C_{i'j'k'l'} - C_{ijkl})\}$. Let $u'_c(t)$ be a deformation onto a *particular* bisector:

$$u'_c(t) = \inf\{u \mid J_0(t, u) \in \ker(C_{i'j'k'l'} - C_{ijkl})\} \tag{5.10}$$

then u'_c is continuous because by definition

$$(C_{ijkl} - C_{i'j'k'l'})(J_0(t, u'_c(t))) = 0 \tag{5.11}$$

and differentiating with respect to t, we obtain

$$\frac{\partial}{\partial t}(C_{ijkl} - C_{i'j'k'l'})(J_0(t, u'_c)) + \left(\frac{\partial u'_c}{\partial t}\right) \frac{\partial}{\partial u'_c}(C_{ijkl} - C_{i'j'k'l'})(J_0(t, u'_c)) = 0 \tag{5.12}$$

which can be rearranged to yield

$$\frac{\partial u'_c}{\partial t} = -\frac{\frac{\partial}{\partial t}(C_{ijkl} - C_{i'j'k'l'})(J_0(t, u'_c))}{\frac{\partial}{\partial u'_c}(C_{ijkl} - C_{i'j'k'l'})(J_0(t, u'_c))} \tag{5.13}$$

and therefore $\frac{\partial}{\partial t}u'_c(t)$ is finite, because the denominator above is non-zero by our general position assumption. Now we observe that $u_c(t)$ can be constructed by pasting together segments of $u'_c(t)$ for various bisectors $\in$ conc(V), and we must show that they agree at their endpoints. The proof is by contradiction. Suppose we had

$$u_c(t) = \begin{cases} u'_c(t) & \text{for } t \in (t_0, t_1]; \\ u''_c(t) & \text{for } t \in (t_1, t_2); \end{cases} \tag{5.14}$$

$$\text{with} \quad u'_c(t_1) \neq u''_c(t_1)$$

for u' and u'' derived from distinct bisectors. Then since u_c is the minimum of all such u, we must have $u'_c(t_1) < u''_c(t_1)$

But this means that as u increases, $J_0(t_1, u)$ crosses two bisectors in conc(V) between C_{ijkl} and other constraints. This is impossible because all constraints C have $|\mathbf{N}(\mathbf{q})| = 1$, and it follows that

$$\frac{\partial}{\partial u} C_{i'j'k'l'}(J_0(t_1, u)) \leq 1 \tag{5.15}$$

$$\frac{\partial}{\partial u} C_{ijkl}(J_0(t_1, u)) = 1$$

That is, all constraints increase no faster than C_{ijkl} with the deformation parameter u. By our general position assumption if $C_{ijkl} \neq C_{i'j'k'l'}$, then the inequality in (5.15) is strict, and so if the clause $C_{i'j'k'l'}$ becomes critical at u'_c, it shares a min node lowest ancestor with C_{ijkl}. Since all constraints in the tree increase more slowly than C_{ijkl}, the value of this min node will be less than the value of C_{ijkl} for $u > u'_c(t_1)$. Therefore C_{ijkl} cannot be critical for any $u > u'_c(t_1)$, contradicting the assumption that $u''_c(t_1)$ is distinct from $u'_c(t_1)$. So all bisectors in conc(V) between C_{ijkl} and other constraints agree at their endpoints, and $u_c(t)$ is continuous by pasting. This shows that J_1 is continuous. □

The homotopy described above is illustrated in figure 5.3. The direction of the gradient of the distance function is shown by the arrows. Each point on the path moves in the gradient direction until it hits a concave bisector.

If $p(0) \in$ conv(V) the situation is more complicated, because the deformation J_1 pushes $p(0, 1]$ away from $p(0)$. To correct this, we first compress the first half of p to a point:

Lemma 5.1.2 *Let $p : I \rightarrow F$ be a path in free space. Then p is homotopic to a path p' such that*

(i) $p'([0, \frac{1}{2}]) = p(0)$

(ii) $p'(1) = p(1)$

(iii) $p'(I) = p(I)$

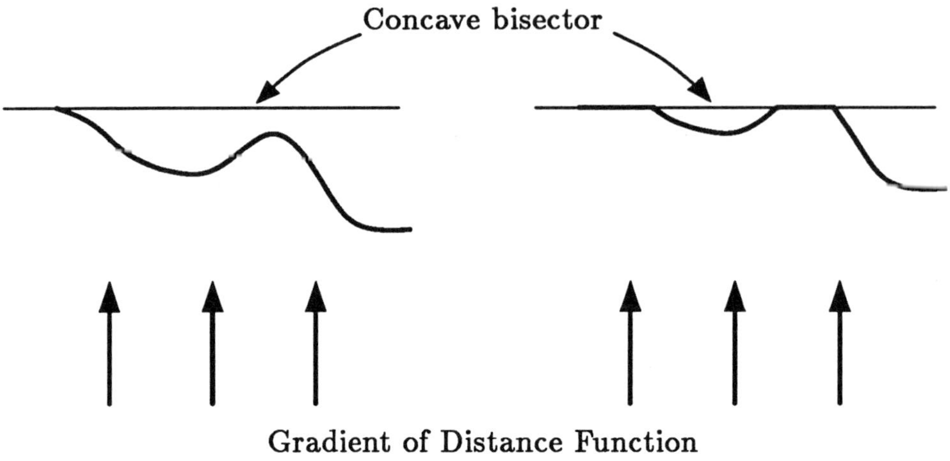

Figure 5.3: Illustration of the homotopy of lemma 4.2

Proof. The required homotopy is

$$J(t,u) = \begin{cases} p(0) & \text{if } t \le \frac{1}{2}u; \\ p\left(\frac{2t-u}{2-u}\right) & \text{otherwise;} \end{cases} \tag{5.16}$$

so $J(t,0) = p(t)$, and $J(t,1) = p'(t)$. □

We apply the homotopy of lemma 5.1.2 to p to give us a path p'. Applying the homotopy J_1 to the path segment $p'|_{[\frac{1}{2},1]}$ continuously deforms this path segment onto conc(V). Then we define a new homotopy which slowly "unravels" $p'|_{[0,\frac{1}{2}]}$ from $p(0)$. All the points in this homotopy have the same orientation, and for each value of the deformation parameter u, the path consists of a finite number of straight line segments.

We now construct a third homotopy. The construction is inductive and we start with a homotopy that gives us two straight line segments. If the orientation at the configuration $p(0)$ is $\mathbf{q}_0$, every point on the joining path will also have orientation $\mathbf{q}_0$. We define two vectors in position space $\mathbf{N}_0$ and $\mathbf{M}_0$ which will be used to define the joining path segment.

Let C_{ijkl} and $C_{i'j'k'l'}$ be the two constraints that are critical at $p(0)$, then $\mathbf{N}_0$ lies in the plane of the bisector of C_{ijkl} and $C_{i'j'k'l'}$. $\mathbf{N}_0$ is normalized so that $\mathbf{N}_0 \cdot \mathbf{N}_{ijkl}(\mathbf{q}_0) = 1$ and it follows that $\mathbf{N}_0 \cdot \mathbf{N}_{i'j'k'l'}(\mathbf{q}_0) = 1$. A second vector $\mathbf{M}_0$ is chosen so that $\mathbf{M}_0 + \mathbf{N}_{ijkl}(\mathbf{q}_0) = \mathbf{N}_0$, and so:

$$\mathbf{N}_0 = \frac{\mathbf{N}_{ijkl} + \mathbf{N}_{i'j'k'l'}}{1 + \mathbf{N}_{ijkl} \cdot \mathbf{N}_{i'j'k'l'}} \tag{5.17}$$

$$\mathbf{M}_0 = \frac{\mathbf{N}_{i'j'k'l'} - (\mathbf{N}_{ijkl} \cdot \mathbf{N}_{i'j'k'l'})\mathbf{N}_{ijkl}}{1 + \mathbf{N}_{ijkl} \cdot \mathbf{N}_{i'j'k'l'}} \tag{5.18}$$

where $\mathbf{N}_{ijkl}$ (shorthand for $\mathbf{N}_{ijkl}(\mathbf{q}_0)$) is the normal vector to the critical constraint C_{ijkl} at orientation $\mathbf{q}_0$, and $\mathbf{N}_{i'j'k'l'} = \mathbf{N}_{i'j'k'l'}(\mathbf{q}_0)$ is the vector normal to the critical constraint $C_{i'j'k'l'}$. (Note that $\mathbf{N}_{ijkl}$ cannot equal $-\mathbf{N}_{i'j'k'l'}$ by our general position assumptions).

We can now define the joining homotopy $J_2 : I \times I \to \Re^3 \times SO(3)$ which deforms the path segment $p'|_{[0,\frac{1}{2}]}$ (with p' reparametrized so that its domain is I):

$$J_2(t,u) = \begin{cases} p(0) + 2t\mathbf{N}_0, & \text{if } t \in [0, \frac{1}{2}u]; \\ p(0) + u\mathbf{N}_0, & \text{if } t \in [\frac{1}{2}u, 1 - \frac{1}{2}u]; \\ p(0) + (2-2t)\mathbf{M}_0 + u\mathbf{N}_{ijkl}(\mathbf{q}_0), & \text{if } t \in [1 - \frac{1}{2}u, 1]; \end{cases} \tag{5.19}$$

The action of this homotopy on a path segment is illustrated in figure 5.4. All points not on a bisector move in the direction of the gradient of the distance function. Points which are on the bisector do not move, except the single corner point. This point may be thought of as a "ball of string", which continuously unravels and allows points on the two straight line segments to move as described above.

Now $\mathbf{N}_0$ lies in the plane of the bisector of the two constraints that are critical at $p(0)$, but it is possible that as u increases, $J_2(\frac{1}{2}, u)$ leaves the convex part of the diagram before reaching the concave part. That is, there may be bends in the convex part of the diagram which must be tracked. We must therefore stop the deformation at this point by defining

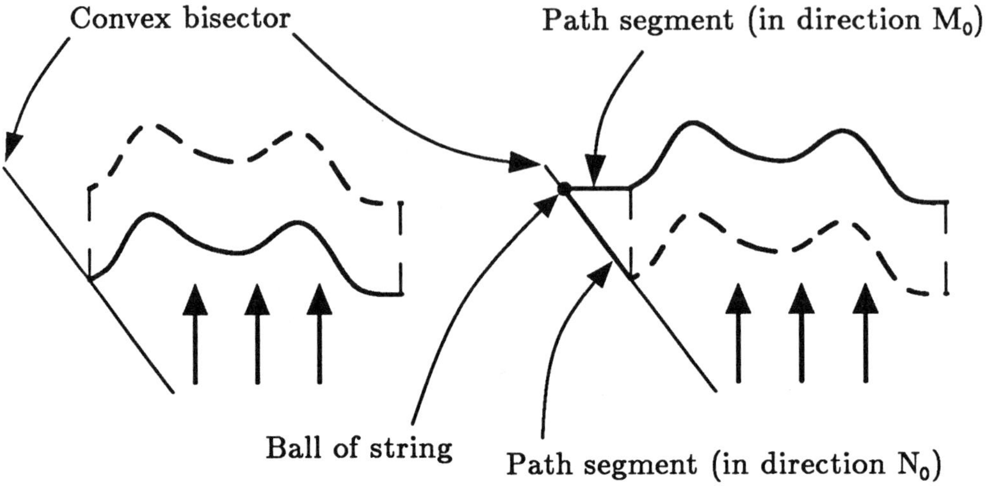

Figure 5.4: Homotopy to continuously link a deforming path to a point on a convex bisector

$$u_v = \sup\{u \mid \; C_{ijkl} \text{ is critical throughout } J_2(\tfrac{1}{2}, [0, u])\} \tag{5.20}$$

and once again we define a homotopy which stops points when they reach the diagram:

$$\begin{aligned} J_3(t, u) &= J_2(t, \min(u, u_v, u_c(t))) \\ \text{where} \quad u_c(t) &= \inf\{\, u \mid \; J_2(t, u) \in \text{conc}(V)\,\} \end{aligned} \tag{5.21}$$

and then we have:

Lemma 5.1.3 *Let* $p : I \to \Re^3 \times SO(3)$ *be a path having* $p(I) = p(0) \in \text{conv}(V)$. *Then* J_3 *is a homotopy of* p *and a path* p' *such that*
(i) $p'(0) = p(0)$
(ii) $p'([0, 1 - \frac{1}{2}u_v]) \subset \text{conv}(V)$
(iii) $p'((1 - \frac{1}{2}u_v, 1]) \cap V \subset \text{conc}(V)$

Proof.

Parts (i) and (ii) follow immediately from the definition of J_2 and J_3. Part (iii) states that points in the interval $(1 - \frac{1}{2}u_v, 1]$ that are mapped into the diagram by p' are mapped into the concave part of the diagram. For this we notice that

$$\frac{\partial J_2(t,u)}{\partial u} = \mathbf{N}_{ijkl}(\mathbf{q}_0) \tag{5.22}$$

for $t > 1 - \frac{1}{2}u$, while $J_2(t,u) \in \text{conv}(V)$ if $t \leq u$. Therefore for $t > 1 - \frac{1}{2}u$ the following is true:

$$\begin{aligned} \frac{\partial C_{i'j'k'l'}(J_2(t_1,u))}{\partial u} &\leq 1 \\ \frac{\partial C_{ijkl}(J_2(t_1,u))}{\partial u} &= 1 \end{aligned} \tag{5.23}$$

So as u increases, all constraints increase no faster than C_{ijkl}. So another constraint can only become critical if its lowest common ancestor with C_{ijkl} is a min node, and such a configuration must be in conc(V).

For continuity of J_3, first we notice that J_2 is continuous, and J_3 will be continuous if $u_c(t)$ defined in (5.21) is continuous. For this we notice that the use of u_v in the min function in (5.21) guarantees that C_{ijkl} is critical at configuration $J_3(t,u)$ for all t and u. The rest of the proof of continuity is identical to the proof of continuity of $u_c(t)$ in lemma 5.1.1, using the rate of change condition in (5.23). □

Lemma 5.1.4 *If $p : I \to F$ is a path having $p([0, \frac{1}{2}]) = \{p(0)\} \subset \text{conv}(V)$ then p is homotopic to a path $p' : I \to V$ such that $p'(0) = p(0)$.*

Proof. The proof is inductive. We define a sequence of partial homotopies, that is, maps $J^n : I \times [u^{n-1}, u^n] \to F$ such that $J^n(t, u^n) = J^{n+1}(t, u^n)$. Then we show that the number m of partial homotopies required is finite. (We employ a superscript notation which will prove convenient in our inductive argument).

Inductive hypothesis. The input to our construction is a path p^n, a value of $u^n \in I$, and points t_0^n and t_1^n in I such that

(i) $p^n(t) \in \text{conv}(V)$ for $t \leq t_0^n$

(ii) $p^n([t_0^n, t_1^n]) = \{p^n(t_0^n)\} \subset \text{conv}(V)$

(iii) C_{ijkl} is critical on $p^n(I)$.

From C_{ijkl} we use lemma 5.1.1 to define a homotopy J_1 for the path segment $p^n|_{[t_1^n,1]}$ (reparametrized to I). Similarly, let $C_{i'j'k'l'}$ be a constraint which is critical at $p^n(t_0^n)$ and whose lowest common ancestor with C_{ijkl} is a max node. For these two constraints, we use lemma 5.1.3 to define an unravelling homotopy J_3 of the path segment $p^n|_{[t_0^n,t_1^n]}$ (reparametrized to I). This gives us a value of u_v as in (5.20), and we define $u^{n+1} = u^n + u_v$. Now J^{n+1} can be defined on the range $u \in [u^n, u^{n+1}]$ as:

$$J^{n+1} = \begin{cases} p^n(t) & \text{for } t \in [0, t_0^n]; \\ J_3\left(\frac{t-t_0^n}{t_1^n-t_0^n}, u - u^n\right) & \text{for } t \in [t_0^n, t_1^n]; \\ J_1\left(\frac{t-t_1^n}{1-t_1^n}, u - u^n\right) & \text{for } t \in [t_1^n, 1]; \end{cases} \tag{5.24}$$

then J^{n+1} is a homotopy because J_3 and J_1 agree on their intersection, and $J_3(0, u) = p^n(t_0^n)$. We can define a new path $p^{n+1}(t) = J^{n+1}(t, u^{n+1})$ and points

$$t_0^{n+1} = t_0^n + \frac{u_v}{2}(t_1^n - t_0^n) \quad \text{and} \quad t_1^{n+1} = t_1^n - \frac{u_v}{2}(t_1^n - t_0^n) \tag{5.25}$$

and it is readily verified that these satisfy the inductive hypothesis.

For the base case, we set $p^0 = p$, $u^0 = 0$ and $t_0^0 = 0$, $t_1^0 = \frac{1}{2}$, which clearly satisfies the inductive hypothesis.

Finiteness. For termination, we must show that after a finite number of steps m, $p^m(t_0^m) \in \text{conc}(V)$. Suppose $p^n(t_0^n) \notin \text{conc}(V)$, and let $C_{i'j'k'l'}$ be the constraint (along with C_{ijkl}) which is critical at $p^n([t_0^{n-1}, t_0^n])$, i.e. these are the constraints used to define the homotopy J_3 for J^n. A third constraint $C_{i''j''k''l''}$ is also critical at $p^n(t_0^n)$, by (5.20). This constraint has a max node lowest common ancestor with $C_{i'j'k'l'}$, which implies that

$$\frac{\partial}{\partial u} C_{i'j'k'l'}(J_3(\tfrac{1}{2}, u)) < \frac{\partial}{\partial u} \bar{C}_{i''j''k''l''}(J_3(\tfrac{1}{2}, u)) \tag{5.26}$$

and from (5.17) it follows that

$$(N_{ijkl} + N_{i'j'k'l'}) \cdot N_{i'j'k'l'} < (N_{ijkl} + N_{i'j'k'l'}) \cdot N_{i''j''k''l''} \tag{5.27}$$

which implies

$$N_{ijkl} \cdot N_{i'j'k'l'} < N_{ijkl} \cdot N_{i''j''k''l''} \tag{5.28}$$

but the condition (5.28) defines a total ordering on the constraints distinct from C_{ijkl}. That is, as u increases and $p^n(t_0^n)$ is deformed according to some J_2, a new constraint $C_{i''j''k''l''}$ can only become critical if it satisfies condition (5.28). Once $C_{i''j''k''l''}$ has become critical, $C_{i'j'k'l'}$ can never again become critical. Thus we need define homotopies J^n at most once for each constraint, and so their number m is bounded by the number of constraints.

We then construct J_1 for the path segment $p^m|_{[t_1^m,1]}$ reparameterized to I. The final homotopy is defined for the range $u \in [u^m, 1]$ as:

$$J^{m+1}(t,u) = \begin{cases} p^m(t) & \text{for } t \in [0, t_1^m]; \\ J_1(\frac{t-t_1^m}{1-t_1^m}, u - u^m) & \text{for } t \in [t_1^m, 1]; \end{cases} \tag{5.29}$$

then the homotopy

$$J_4(t,u) = J^n(t,u) \quad \text{with} \quad u \in [u^{n-1}, u^n] \quad \text{for} \quad n = 1, \ldots m+1 \tag{5.30}$$

is continuous, and defines a homotopy between $p(t) = J_4(t,0)$ and a path $p'(t) = J_4(t,1)$ such that $p'(I) \subset V$. □

Theorem 5.1.5 *Let $p : I \to F$ be a path with endpoints in V. Then p is path homotopic in F to a path p' with the same endpoints which lies entirely in V.*

Proof. We first apply the homotopy of lemma 5.1.2 to all path segments $p|_{[t_n,t_{n+1}]}$ with an endpoint in conv(V). This does not displace endpoints in V. Then we construct a global homotopy J by pasting together homotopies J_1 for path segments with an endpoint in conc(V), and homotopies J_4 for the remaining path segments. The resulting homotopy is continuous if these homotopies agree at their endpoints. Firstly, both J_1 and J_4 to not displace endpoints in V. Therefore they agree at endpoints in V. At free endpoints both satisfy the *free endpoint condition*: Assume that after reparametrization, $p(1)$ is a free endpoint. Then

$$\begin{aligned} J_n(1,u) &= p(1) + \min(u, u_c)\mathbf{N}_{ijkl} \qquad \text{for} \qquad n = 1,4 \\ \text{with} \qquad u_c &= \inf\{u \mid p(1) + u\mathbf{N}_{ijkl} \in \mathrm{conc}(V)\} \end{aligned} \tag{5.31}$$

thus J is continuous, and we define $p'(t) = J(t,1)$. Since $J_1(I,1) \subset V$ and $J_4(I,1) \subset V$, we have $p'(I) \subset V$. $\square$

Finally, suppose that in a motion planning problem we are given a start configuration $(\mathbf{x}, \mathbf{q})$ which is not on V. Then exactly one constraint C_{ijkl} is critical there. We apply the homotopy J_1 to the constant path at $(\mathbf{x}, \mathbf{q})$ to attain the diagram; that is, we plan a straight-line path in direction $\mathbf{N}_{ijkl}(\mathbf{q})$ to reach V from the start.

5.2 Complexity bounds

We have given a definition of the simplified Voronoi diagram V in the configuration space of a polyhedron in 3-space. This definition does not constitute an algorithm, so our bounds depend on the algorithm used with the diagram. We assume that the diagram will be used as input to a version of the roadmap algorithm of chapter 4. That algorithm computes one-dimensional skeletons of semi-algebraic sets in time $(d^{O(r^2)} n^r \log n)$ for a semi algebraic set defined by n polynomials of degree d in r variables. In our case the number of variables and the degree of the equations are constants.

A naive bound on the complexity of computing a skeleton of V would be $O(n^{12} \log n)$ if we are given n constraints, because the diagram is a subset of the zero sets of all $O(n^2)$ bisectors of constraints. This bound can be reduced to $O(n^7 \log n)$ by noticing that the diagram has a simple stratification (decomposition into a union of disjoint manifolds). The diagram is a subset of the set of all m-sectors, where an m-sector is the set of points where m constraints have the same value. If the constraints are in general position, each m-sector is a manifold of codimension $m-1$. There are $O(n^m)$ m-sectors, and by the codimension condition, m must be less than or equal to 7. The complexity of computing the skeleton of this stratification is $O(n^7 \log n)$.

While its worst case bounds are poor, the actual performance of the algorithm is expected to be much better, because V approximates the euclidean Voronoi diagram, as shown in figures 5.1 and 5.2. The evidence for this is that the complexity of the euclidean Voronoi diagram for a set of n points in r dimensions is $O(n^{\lfloor \frac{(r+1)}{2} \rfloor})$, and the euclidean Voronoi diagram for disjoint line segments in the plane has linear size.

This conjecture is supported by some experimental evidence. We have implemented an algorithm for constructing the simplified Voronoi diagram for the following configuration spaces: $\Re^2$, the case of an arbitrary polygon translating in the plane amidst polygonal obstacles, and $\Re^2 \times S^1$, which allows the moving polygon to rotate as well as translate. In many cases the size of V has been observed to remain roughly linear, as in figure 5.1, which the implementation produced.

Chapter 6

Lower Bounds for Motion Planning

In this chapter [1] we give new lower bounds for several basic variations on the movers' problem. Our results apply to problems in which the number of degrees of freedom is small and fixed (two or three). Motion planning for these environments is solvable in polynomial time if there are no auxiliary constraints. However, if a shortest path is sought, or if velocity bounds or uncertainty are added the planning problem is provably hard. Our proofs are based on a technique called *free path encoding*. In free path encoding we represent discrete state using equivalence classes of paths. These path classes have "free" boundaries, which means that different path classes are not separated by obstacle surfaces. This allows us to generate and manipulate a number of path classes which grows exponentially with the complexity of the obstacles. In the first two proofs of NP-hardness of shortest path and dynamic motion planning, each path class represents a satisfying assignment to a set of boolean variables. In the third proof, non-deterministic exponential time hardness of compliant motion planning with uncertainty, we use path classes to represent the contents of a exponential-length Turing machine tape.

In section 6.1 we give a brief description of the free path encoding technique for NP-hardness proofs, and describe the main structures required.

[1] Parts of this chapter describe joint work with Prof. John Reif of Duke University

We first derive lower bounds on the number of path classes for single-source multiple-destination problems. Then we describe the filtering scheme that allows us to encode a satisfying assignment to a 3-SAT formula in the shortest path. The environment we construct is an approximation to an ideal environment which has slits of zero thickness, so we must verify that the behaviour of paths in the actual environment is sufficiently close to the idealisation. Finally, we verify that the number of bits required to describe all the dimensions in the environment is polynomial in the size of the 3-SAT formula, and that a description of the environment can be computed in polynomial time.

Section 6.2 gives a similar proof of NP-hardness for two-dimensional dynamic motion planning with velocity limits (the "asteroid avoidance" problem). The basic structures used in the proof are the same as for the shortest path proof, but they have different physical realisations. Once again, the actual behaviour of the environment approximates the behaviour of an ideal environment, and we need to find the accuracy required for this approximation to be valid. Finally, we must verify that a description of such an environment can be computed in polynomial time.

Section 6.3 covers compliant motion planning with uncertainty under the generalized damper dynamic model described in the first chapter. Even verify one step of such a plan is shown to be NP-hard. We then describe a class of polyhedral environments for which *finding* a guaranteed motion plan is non-deterministic exponential-time hard. The proof differs from the previous two proofs in that the encoding is "dynamic". Free path classes may be written or erased, and so may be used to directly encode a Turing machine tape. Because of uncertainty however, this tape decays with time, and this limits the duration of the simulation of the Turing machine.

6.1 Lower Bounds for the Shortest Path Problem

6.1.1 Free Path Encoding for Shortest Paths

First we define the *shortest route* from any point x in space to a source q as the sequence of environment edges traversed by a shortest path from x to q. In some cases there may be more than one shortest route from x to q. We define an equivalence relation such that two points are equivalent if they

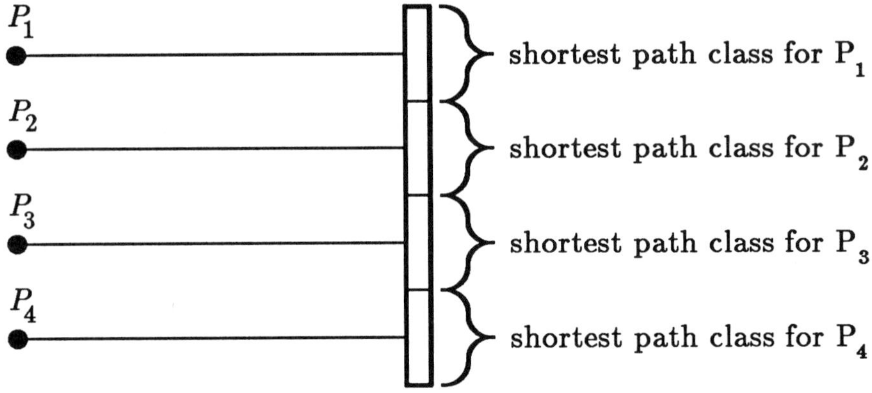

Figure 6.1: Path classes and virtual sources for a single slit.

have the same set of shortest routes to q. This equivalence relation gives us a partition of free space and allows us to speak of *shortest path classes.*

To simplify things we will only examine these classes at certain "one dimensional" slits. These slits actually have some finite width ϵ and lie in horizontal plates of thickness also ϵ. In our construction it is possible to represent the path classes within a slit using the concept of a "virtual source". A virtual source is a point p_i such that the actual distance from a point x in the path class to the start point is the same (to within a small additive error) as the straight line distance from x to p_i. The straight line is often referred to as an "unfolding" of the actual path. Each shortest path class has its own virtual source, and the sources will be evenly spaced and equidistant from the slit as shown in figure 6.1.

6.1.2 The Environment

We will be generating (2^n) path classes in our construction, and each class may be thought of as encoding an n-bit binary string, $b_1, \ldots b_n$. The construction may be broken down into three types of substructure:

- Path Splitter: This doubles the number of shortest path classes by splitting them.

- Path Shuffler: Performs a perfect shuffle of path classes.
- Literal Filter: Filters for those paths which have a particular bit equal to zero or one in their encoding.

Path Splitter

The path splitter has the property that if its input slit contains n shortest path classes as described above, its output slit will contain $2n$ path classes. A splitter consists of a stack of 3 horizontal plates separated by ϵ, and is shown from above in figure 6.2. The top plate covers the x-y plane except for the input slit S_{in}. The middle plate has two slits at 45°, labelled S_1 and S_2, and the bottom plate contains the output slit S_{out}. Any shortest path from S_{in} to S_{out} passes through either S_1 or S_2.

The splitter uses the unfolding of paths to generate many virtual sources. The principle of path unfolding is illustrated in figure 6.2. For any point p in the output slit S_{out}, the shortest path from that point to the virtual source P_1 consists of two straight line segments which make equal angles with S_2 (actually this is only true in the limit as $\epsilon \to 0$, but we will deal the non-zero ϵ case shortly). By mirror symmetry there is a second virtual source P_1' such that the distance of the shortest path from it any point p in S_{out} to P_1 is the same as the *straight line distance* from p to P_1'.

Suppose we have n virtual sources for the n shortest path classes in S_{in}. We have seen that S_2 behaves like a mirror and reflects each virtual source P_i to a new position P_i'. Since S_1 also produces a reflected copies P_i'' of the P_i, S_{out} now sees $2n$ virtual sources, as shown in figure 6.3. In other words there are twice as many shortest path classes in S_{out} because each path class from S_{in} bifurcates into a class that passes through S_1 and a class that passes through S_2. The spacing between virtual sources remains the same, so the output slit of a splitter is twice the length of its input slit.

The geometric argument above is only true if the plates have zero thickness and slits have zero width, however it does give valid lower bounds for a finite thickness, finite width environment. In section 6.1.4, we derive upper bounds which differ by an additive multiple of ϵ. With these bounds and a sufficiently small ϵ, we can guarantee that path classes remain distinct,

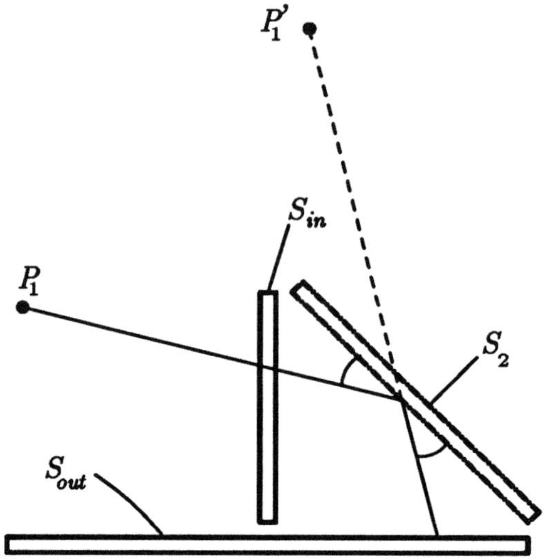

Figure 6.2: Path splitter, showing the principle of path unfolding.

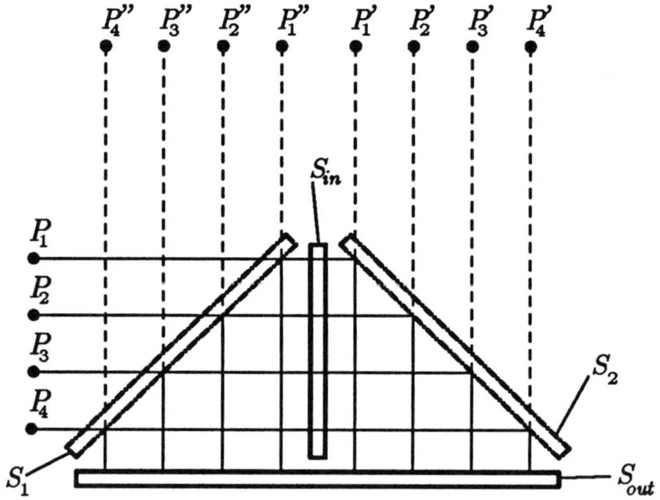

Figure 6.3: Path splitter showing doubling of the number of path classes.

and in section 6.1.5, we show that the three-dimensional environment has a polynomial size description.

Path Shuffler

A shuffler is shown from above in figure 6.4. It consists of 4 horizontal plates of width and spacing ϵ. The top two plates simply split the virtual sources into two groups, and half the sources appear in each of the output slits of the second plate, see figure 6.4(a). The second, third and fourth plates behave rather like a twisted splitter. The second plate contains the two slits S_1 and S_2, the third plate contains the diagonal slits S_3 and S_4, and the fourth plate contains the output slit S_{out}. Paths from S_1 to S_{out} are constrained by a barrier to pass through S_3. Thus S_3 acts as a mirror and produces a copies of the virtual sources in S_1. S_4 produces a similar copies of the sources in S_2, however these two images are displaced by half the source spacing δt. This has the effect of interleaving the two sets of path classes, so the class numbers of the lower half ($b_1 = 0$) of the input slit are doubled, while the path numbers of the upper half ($b_1 = 1$) are doubled and incremented. This corresponds to a left shift of the encoding with wrap around of the leading bit, i.e. a circular left-shift. The spacing between virtual sources is halved, so the output slit of a shuffler has half the length of its input slit.

Literal Filter

The literal filter is designed to filter for those path classes whose encodings have a zero or a one in the ith bit. It does this by stretching all other path classes. Recall that the jth path class encodes a binary string $b_1, \ldots b_n$. To filter for paths having $b_i = 0$, we put a barrier in the way of paths having $b_i = 1$, thus forcing the shortest paths from these sources to be stretched as they pass around the barrier. The problem with doing this on the original encoding is that the path classes we want to filter are not adjacent, and the barrier may need to be split into exponentially many sections, e.g. for b_n we need to cover every second path class. Instead we use shufflers to rotate the encoding so that the ith bit becomes the most significant bit of the encoding, at which point all paths having $b_i = 1$ are in the same half of the slit.

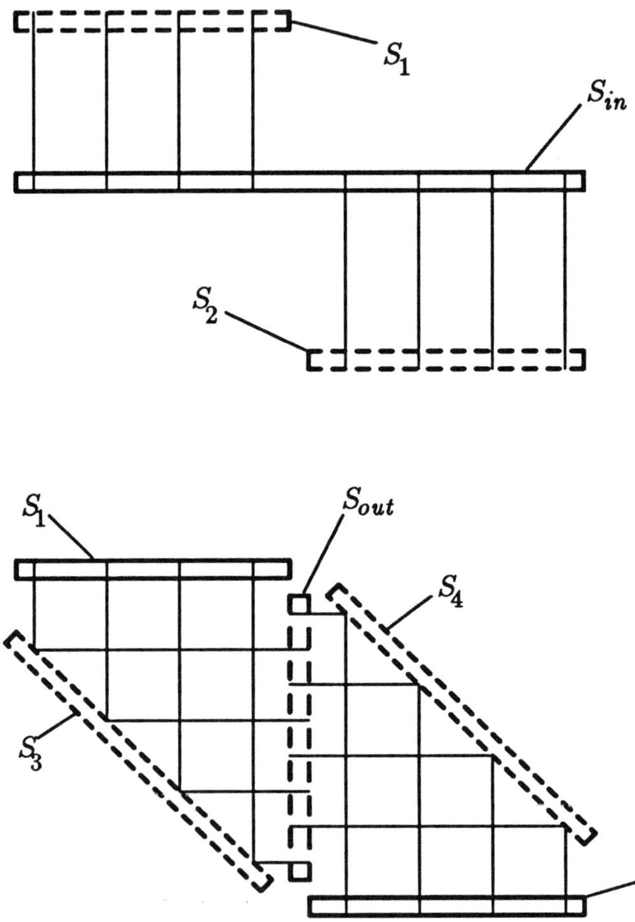

Figure 6.4: Path shuffler. The first two plates are shown at the top, and plates two, three, and four at the bottom of the figure.

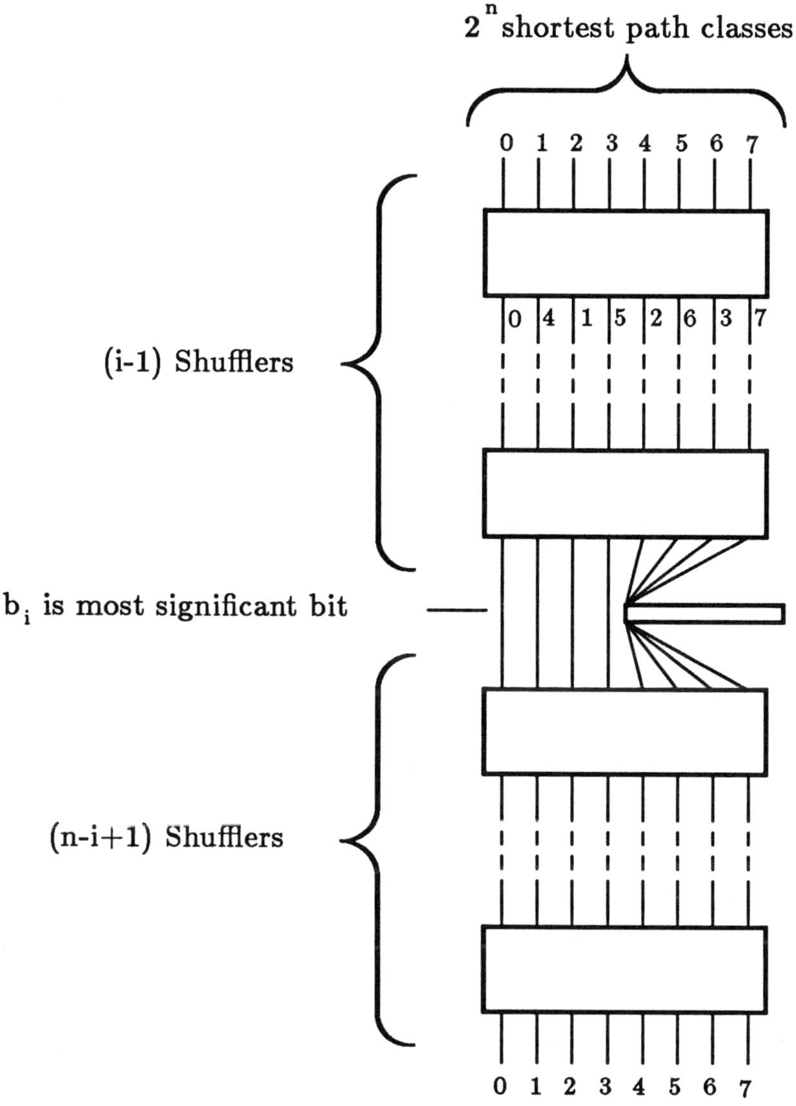

Figure 6.5: A literal filter with barrier to stretch paths having $b_i = 1$.

A literal filter for $b_i = 0$ is shown in figure 6.5. Each shuffler performs a 1-bit left shift of the encoding, and after cascading $(i-1)$ shufflers, the encoding becomes $b_i, \ldots b_n, b_1, \ldots b_{i-1}$. Once the ith bit has become the most significant bit, we block off the upper half of the slit. We then add $(n-i+1)$ more shufflers to restore the original encoding. Now the paths which have $b_i = 1$ will have been stretched slightly by having to travel around the barrier. In section 6.1.4, equation (6.11) we give bounds on the amount of stretching necessary, and show that stretched paths can be distinguished from unstretched ones.

6.1.3 Lower Bounds for 3-d Shortest Path

The following theorem follows immediately from the properties of splitters:

Theorem 6.1.1 *The number of shortest path classes in a polyhedral environment may be exponential in the number of faces, edges and vertices in the environment, and $\Omega(2^{\sqrt{N}})$ where N is the length of the description of the environment.*

Proof We cascade n splitters each of which has constant number of faces, edges and vertices, giving us 2^n shortest path classes. Applying the results of section 6.1.5, we need only $O(n)$ bits to describe each environment vertex (since there are no shufflers), and thus $N = O(n^2)$ bits for the whole environment. $\square$

Theorem 6.1.2 *The problem of finding a shortest path under any L^p metric in a three-dimensional polyhedral environment is NP-hard.*

Proof Given a 3-SAT formula, we construct a polyhedral environment, and source and target points, such that knowledge of the shortest path between source and target allows us to decide in polynomial time if the formula is satisfiable. The size of the environment description is polynomial in the formula size.

Recall that a 3-SAT formula in n variables $b_1, \ldots b_n$ has the form

$$\bigwedge_{i=1,\ldots m} C_i \tag{6.1}$$

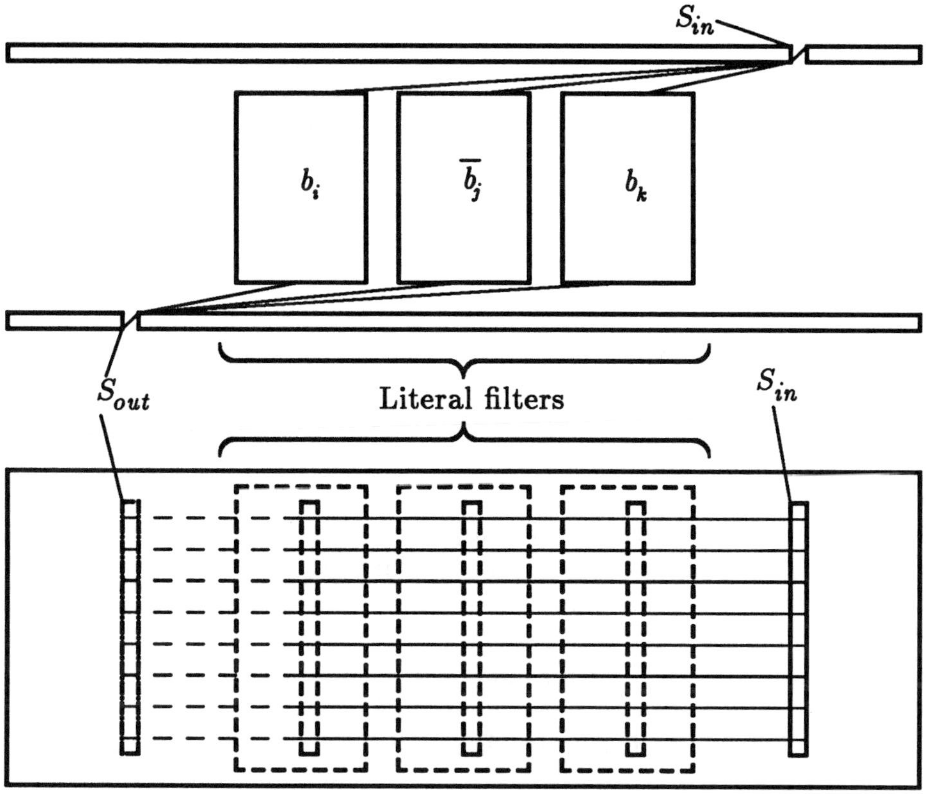

Figure 6.6: A clause filter.

where each C_i is a clause of the form $(l_{i1} \vee l_{i2} \vee l_{i3})$. Each literal l_{ij} in turn is either a variable or the negation of a variable. The size of the formula can be measured as the sum of the number of variables n and the number of clauses m.

First we cascade n splitters below the source point, to give us 2^n shortest path classes. Each path encodes an assignment to the n variables through the binary representation of the path number. Then we progressively filter out those classes whose encodings do not satisfy a particular clause in the formula.

To filter for a particular clause, we place 3 literal filters in parallel (each consisting of n shufflers) as shown in figure 6.6. This structure implements a disjunction of literals because if an assignment satisfies any of the three literals there will be a short path through the corresponding literal filter. The collection of literal filters and top and bottom plate will be called a clause filter. We can cascade clause filters to represent all the clauses in the 3-SAT formula, and the output slit of the mth clause filter will contain short path classes only for those assignments (if any) which satisfy the formula.

The final step is to collect all these path classes into a single path class. This is done using a series of n inverted splitters. The effect is to form the disjunction of all the satisfying assignments. In the absence of any barriers in the preceding clause filters, there would be a single path class in the output slit and a single virtual source. Below this slit and aligned with the source, we place the target point q'. Let l be the approximate length of the shortest path from q to q' when the barriers are removed. We then ask if the shortest path from q to q' in an environment containing barriers has length close to l. It will if and only if there is some path through the environment which was not stretched by having to go around a barrier. Such a path encodes a satisfying assignment to the formula, and since we have encoded all possible assignments, the path has length close to l (see section 6.1.4 for a precise definition) if and only if the formula is satisfiable. Finally, by the results of section 6.1.5, the description of the environment has size polynomial in n and m and can be computed in polynomial time. □

Corollary 6.1.3 *Determining even the sequence of edges touched by the shortest path (under an L^p metric) is NP-hard.*

Proof Each path class is uniquely determined by the sequence of splitter edges it touches. This sequence of edges encodes an assignment to the boolean variables. The shortest path will always be an unstretched one, if there are any unstretched paths. So the variable assignment encoded by the shortest path edge sequence is a satisfying assignment, if there is one. This assignment can be substituted into the SAT formula, and it will satisfy the formula if and only if the formula is satisfiable.

Corollary 6.1.4 *For a polyhedral environment of size N, determining $O(\sqrt{N})$ bits of the length of the shortest path between two points in the* *environment is NP-hard.*

Proof. We first construct an environment that encodes a 3-SAT formula of length $\sqrt[4]{N}$ so that m and n are both less than $\sqrt[4]{N}$. Using the values computed in section 6.1.5, we notice that the value of ϵ defined in (6.12) leads to a gap between the upper bound on unstretched length and the lower bound on stretched length. This gap is at least $2^{-2nm-3n-3}$, so that $O(nm) = O(\sqrt{N})$ bits suffice to distinguish stretched and unstretched paths. The shortest path length will be the unstretched length if and only if the formula is satisfiable. □

6.1.4 The Virtual Source Approximation

Here we show that an approximate virtual source can be used to accurately model the path lengths at the input and output of a shortest path splitter or shuffler. We show that the error at the output of a shortest path splitter is 5ϵ greater than the error at its input. The proof generalizes to shufflers which have four plates rather than three and gives a bound of 7ϵ.

First we define a local coordinate system for each slit, fixed at one end of the slit (say the end closest to paths which encode 0). Let the t coordinate measure position along the slit. Thus t lies in the range $[0, l]$ where l is the length of the slit. Let u, and v measure respectively the horizontal and vertical position of a point in the slit, so that u and v both lie in the range $[0, \epsilon]$. If u increases in the upward direction, the position of the coordinate origin in the slit is completely determined when we add the constraint that the t-u-v system be right-handed.

Let $d^i(x, q)$ denote the distance of the shortest path in the ith path class from a point $x = (t, u, v)$ in the slit to the actual source q. Let $\hat{d}(x, p_i)$ be the straight-line distance from x to the ith virtual source p_i.

Lemma 6.1.5 *The following bound holds at the output of the nth cascaded splitter or half-shuffler:*

$$\hat{d}(x, p_i) \leq d^i(x, q) \leq \hat{d}(x, p_i) + (5n + 2)\epsilon \tag{6.2}$$

Furthermore, $\hat{d}(x, p_i)$ is independent of the u and v coordinates, i.e. we can write $\hat{d}(x, p_i) = \hat{d}(t, p_i)$. In this way we can think of the slit as being one-dimensional, and absorb the values of the u and v coordinates in the error term. The virtual source p_i is specified by two values, its t coordinate $t_i = i\delta t$, assuming sources are uniformly spaced by δt, and a distance h normal to the t axis, (h is the same for all sources in a given slit, so we drop the subscript i). Thus we have a simple expression for $\hat{d}(x, p_i)$:

$$\hat{d}(x, p_i) = ||(t_i - t, h, 0)||_p \tag{6.3}$$

where $x = (t, u, v)$ as before, and $||\cdot||_p$ is the L^p norm defined in (1.2). Our proof of the output hypothesis will be essentially the same for both splitters and half-shufflers, since we only need to consider one edge of the middle plate.

Base case:

If $n = 0$ there is only the true source point q above the center of the first input slit. Let h be the height of the source above the slit. Then clearly $\hat{d}(x, p_i)$ from (6.3) is a lower bound on the distance to any point in the slit, where $p_i = q$ and t_i is half the length of the slit. On the other hand, $\hat{d}(x, q) + 2\epsilon$ is a valid upper bound because every point in the slit is within 2ϵ (under any L^p metric) of a point at the top of the slit whose distance is $\hat{d}(x, q)$ from the source.

Inductive step:

We assume the hypothesis holds at the output of the nth splitter, and consider the path from input to output of the $(n+1)^{st}$. We first obtain a lower bound on the path length between any a point x in the input slit and a point x' in the output slit. For lower bounds, we consider the projection of paths in the x-y plane, since these lengths are less than or equal to the three-dimesional lengths.

For $\epsilon = 0$, let x'' be the point at which the path between x and x' touches the middle plate. Then the total length of the path from the source to x' is bounded below by

$$\hat{d}(x, p_i) + ||x'' - x||_p + ||x', x''||_p \tag{6.4}$$

For each such path, we form a second path by reflecting the first two path segments about the middle slit, similar to figure 6.2. The mirror image path will have the same length as the original under any L^p metric if the middle slit is either parallel to the x or y axis, or at 45° to these axes. We can always build our environment so that this is the case. The length of the three segment path from x' to the mirror image p_i' of the virtual source p_i is bounded below by the straight line distance from x' to p_i' (this follows from the triangle inequality, which holds for any metric). Thus we have

$$\hat{d}(x', p_i') \geq ||(t' - t_i', h + l, 0)||_p \tag{6.5}$$

where $x' = (t', u', v')$ is given in local slit coordinates.

We can also obtain an upper bound on the length of the shortest path between x and x'. To do this we build a path which consists of the horizontal straight line segments in the path for $\bar{\epsilon} = 0$, plus three vertical jumps of ϵ. This path moves vertically from the input slit to the top of the middle plate, horizontally to the to edge of the middle plate, vertically to the bottom edge of the middle plate, horizontally along the bottom surface of the middle plate, and vertically down to the output slit. Finally we need an upper bound which is valid throughout the bottom slit, which may involve traversing the slit in the u and v directions, an extra distance of at most 2ϵ, giving us a total of 5ϵ more than the lower bound for the distance between x and x'. Thus if the upper bound on path length at the input is $\hat{d}(x, p_i) + (5n + 2)\epsilon$, the upper bound at the output is $\hat{d}(x', p_i') + (5(n + 1) + 2)\epsilon$. □

6.1.5 Environment Size

To complete our construction, we must verify that the environment we have defined has a polynomial length description. In particular, we must find the constraints on ϵ for the virtual source approximation to be valid, and ensure that it requires only polynomially many bits. We show furthermore that:

Lemma 6.1.6 *Every environment dimension can be described with $O(nm)$ bits.*

Proof. We first observe that the output slit of a splitter has twice the length of its input slit. We set the length of the top slit (arbitrarily) to be 1. Then after n splitters, we will have a slit of length 2^n with unit spacing between virtual sources. This is the maximum length of any structure in the environment. So we have

Remark 6.1.7 *All dimensions in the environment are $O(2^n)$.*

A shuffler's output slit has half the length of its input slit. Each clause filter consists of n shufflers in cascade, and there are m clause filters in series, so at the output of the last filter we have a slit of length $2^{n(1-m)}$ with virtual sources spaced by 2^{-nm}. All the slits in between have length and spacing in between these two extremes.

When a barrier blocks off half of a particular slit, every path that previously passed through the barrier is displaced horizontally by at least half the source spacing at that slit. For the L^1 metric we assume that the input and output slits of all stages are aligned with the x or y axis, then the new path has length

$$l_{stretched} \geq l + \delta_{min} \qquad \text{for } p = 1 \tag{6.6}$$

where l is the lower bound on the unstretched length of the path from source to target, and δ_{min} is the minimum source spacing. For all other L^p metrics, we align the input and output slits at 45° to the x and y axes. Recall that "mirror image" paths have the same length in either case. Then the length of the shortest stretched path is bounded below by

$$l_{stretched} \geq 2^{-\frac{1}{p}} \sqrt[p]{|l + \delta_{min}|^p + |l - \delta_{min}|^p} \qquad \text{for } 1 < p \leq \infty \tag{6.7}$$

where l is the lower bound, this time under the L^p metric, on the path length. For $p = \infty$, we take the limit of the above expression as $p \to \infty$. Now l is the sum of all the slit lengths plus the distance to source and target points, which we can adjust so that $l = 2^{3n}$. So long as $\delta_{min} \leq 1$, the following lower bound holds for all metrics:

$$l_{stretched} \geq l + \frac{\delta_{min}^2}{4l} \qquad \text{for } 1 \leq p \leq \infty \tag{6.8}$$

To obtain this bound, we raise the right-hand sides of equations (6.7) and (6.8) to the pth power. For $i = 1, \ldots, \lfloor \frac{p}{2} \rfloor$, we observe that the $2i^{th}$ degree term (in δ_{min}) that appears in the pth power of (6.7) is larger than the sum of the $2i^{th}$ and $2(p-i+1)^{th}$ degree terms from the pth power of (6.8). If p is even, all terms are accounted for this way. If p is odd, this pairing leaves an unmatched term in the expansion of (6.8), but it can be added to the sum of the $(p-1)^{th}$ and $(p+3)^{th}$ degree terms, and this sum will still be less than the $(p-1)^{th}$ degree term of (6.7).

Since our proof requires that we can decide whether a path has been stretched, we must have the difference between stretched and unstretched lengths greater than the error in our source approximation. Now the actual unstretched length $l_{unstretched}$ is subject to the following bound (for all metrics):

$$l \leq l_{unstretched} \leq l + (7nm + 10n + 2m + 4)\epsilon \tag{6.9}$$

since there are nm shufflers in cascade with $2n$ splitters each adding 7ϵ and 5ϵ error respectively, and 2ϵ error for the source, target and clause filter plates. To be able to detect a stretched path, we must have the lower bound on stretched length greater than the upper bound on unstretched length, i.e.

$$l_{stretched} \geq l + \frac{\delta_{min}^2}{4l} > l + (7nm + 10n + 2m + 4)\epsilon \geq l_{unstretched} \tag{6.10}$$

and therefore

$$\frac{\delta_{min}^2}{4l} > (7nm + 10n + 2m + 4)\epsilon \tag{6.11}$$

and using the fact that $\delta_{min} = 2^{-nm}$ and $l = 2^{3n}$, the above inequality holds if we set:

$$\epsilon = \frac{2^{-2nm-3n-3}}{(7nm + 10n + 2m + 4)} \tag{6.12}$$

and then ϵ can be specified with $O(nm)$ bits. The entire environment fits in a cube of size $l = 2^{3n}$, and the smallest dimensions that need to be specified are of size approximately ϵ, thus the specification of any dimension in the environment requires $O(nm)$ bits. □

Corollary 6.1.8 *The entire environment can be described with $O(N^4)$ bits, where N is the length of the input SAT formula.*

Proof The environment consists of $3m$ shufflers, and $2n$ splitters. Each shuffler consists of $O(n)$ plates with a fixed number of edges and vertices. Each splitter has a fixed number of faces, edges and vertices. Thus there are $O(nm)$ edges, vertices and faces in the environment each of which can be specified with $O(nm)$ bits, and since both n and m are linear in the formula length N, our construction can be specified in space $O(N^4)$, and computed in polynomial time. □

6.2 Lower Bounds for Dynamic Motion Planning

We can generalize the above technique to motion planning with moving obstacles, where the robot has velocity limits. We have

Theorem 6.2.1 *The 2-d asteroid avoidance problem is NP-hard.*

Proof The proof follows from the results of the previous section if we can construct splitters and shufflers and specify their dimensions with a polynomial number of bits. A single construction can be adapted to implement both functions, and is shown in figures 6.7 and 6.8. We assume the velocity limit is 1. A free path class in this problem is a homotopy equivalence class of *trajectories*. That is, a set of time-parametrized paths such that each one can be obtained from the other by continuous deformation, without collision with obstacles. If a path class at a certain instant of time is a point, then as time goes on, the set of points reachable from that point expands as a circular wavefront.

Figure 6.7 shows a row of path classes which are expanding with time. There is a pair of obstacles moving almost horizontally across the environment. The component of the velocities of these obstacles *normal to their horizontal edges* is 1. So as the path classes expand with time at velocity 1, a small sliver of each path class manages to keep just clear of the obstacles. The obstacles effectively form two copies of the path classes, one copy moving up and the other moving down with velocity 1. Thus this pair of obstacles functions as the first stage of a splitter, and gives us two copies of the original set of path classes.

Figure 6.8 shows the situation some time later. The obstacle A in the figure is one of the obstacles from figure 6.7 with a copy of the path classes

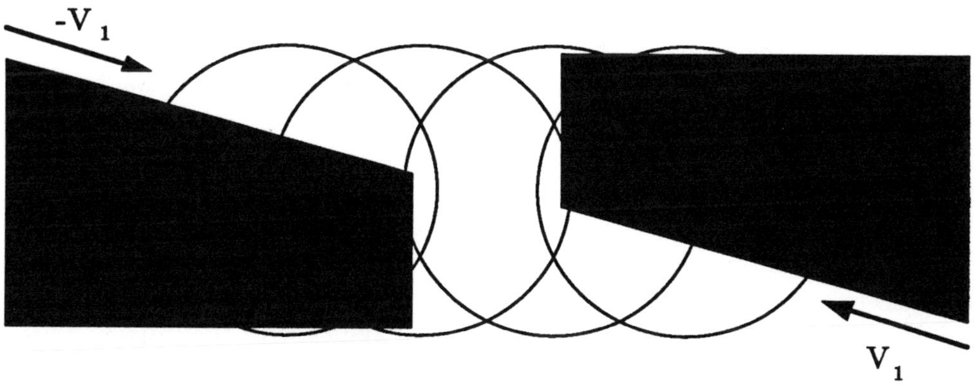

Figure 6.7: Path splitter in an asteroid environment.

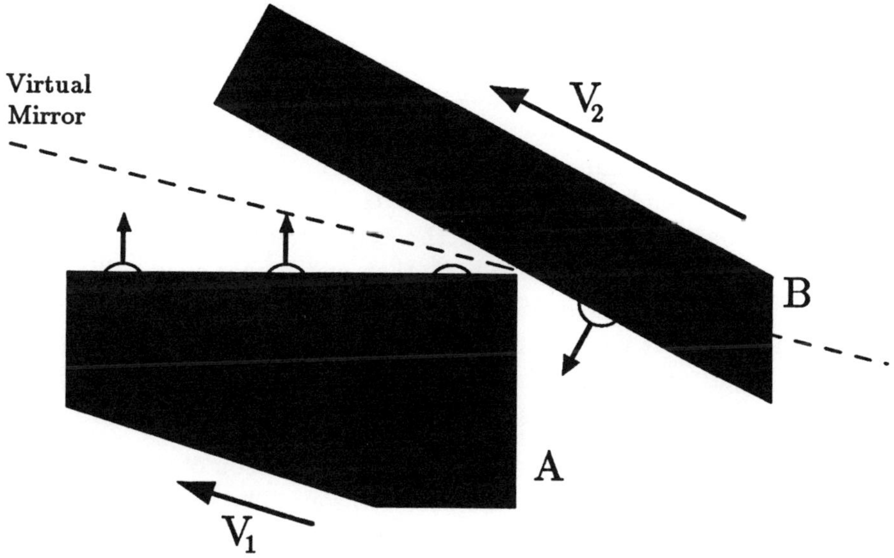

Figure 6.8: Path "reflection" for the asteroid problem. Path classes on object B move normal to B at the maximum velocity.

on its top edge. There is also a new obstacle B, which is moving very fast almost parallel to its lower edge, but nevertheless the component of this velocity normal to the lower edge is still 1. The dotted "virtual mirror" in the figure is the locus of the corner of A. As the obstacle B moves, its lower edge stays a fixed small distance from this corner of A. Each path class is pushed straight up by obstacle A until it "falls off" the right corner of A due to A's motion to the left. As soon as this happens, it will be pushed downward and slightly to the left at unit velocity by obstacle B. A viewer who could only see the path classes and not the obstacles would see each path class move vertically upward until it reached the virtual mirror, and then immediately start moving downward and to the left, such that its initial and final velocity directions make equal angles with the mirror. This is because the upper edge of A and the lower edge of B make equal angles to the virtual mirror.

This situation occurs simultaneously for both sets of path classes from figure 6.7, and after some time all of the path classes have fallen off the A obstacles and are being pushed toward each other (but slightly displaced, because the mirrors are not normal to the path class trajectories) by B obstacles. Assuming the B obstacle have pushed them in opposite directions, the two copies of the path classes are pushed together and eventually lie along a common line normal to the motions of the B obstacles, just before the B obstacles are ready to collide. But the B obstacles do not collide and destroy the path classes. Instead the component of velocity of the B obstacles normal to their edges is so high that just before they collide, they move off to the left and right while leaving small slivers of the path classes intact (a single constant velocity of the B obstacles is carefully chosen that satisfies the normal and tangential constraints). Thus the path classes are left lying along a single line, with some relative displacement along that line.

This displacement depends on the directions of motion of the B obstacles and the distance travelled during the first motion. The displacement may be chosen to implement either a split or a shuffle. For a split, the two sets of path classes are pushed together so that they do not overlap, but lie side by side along the common line. For a shuffle, the displacement is equal to half the width of a set of path classes plus half the spacing between adjacent path classes. So the upper half of one set of path classes is interleaved with the lower half of the other set. This still leaves the two remaining halves on either side of the interleaved paths, and two fast-moving vertical obstacles are used to destroy them.

If the shuffler is part of a literal filter, a fast moving vertical obstacle can be used to destroy one half of the paths. Soon after these vertical obstacles have disappeared, and before the new row of path classes has expanded very much, a new set of A obstacles arrives on the scene to implement the next split or merge. Clause filters are implemented by using the splitting operation of figure 6.7 twice to produce three copies of the path classes, and after literal filtering each copy, the three copies are merged (OR-ed) using fast B obstacles, this time with no displacement. After all the obstacles corresponding to all the clauses of a 3-SAT formula have moved through the environment, any path classes that remain encode a satisfying assignment to the formula. Thus determining if an obstacle-avoiding trajectory exists in this environment is NP-hard.

Quantitatively, we suppose that each path class after the ith shuffle is contained in a rectangle of height h_i and width w_i, and that it contains a circle of diameter h_i. Then if the angle between V_1 and the x-axis is θ_i, and the distance traveled in the first motion is l, we define $h'_i = g$ where g is the gap between the corner of A and the lower edge of B in figure 6.8, and

$$w'_i = (w_i + \sqrt{lh_i})\cos 2\theta_i + h_i \sin 2\theta_i \tag{6.13}$$

as the dimensions of an approximate bounding rectangle (aligned with B) at the end of the first motion. Then the path class at the completion of the second motion is contained in a rectangle of size $h_{i+1} = h_i$ and

$$w_{i+1} \leq w'_i + \sqrt{lg} \tag{6.14}$$

Let n be the number of variables and m the number of clauses in a 3-SAT formula which we wish to encode. Then by choosing $\sin\theta_i \approx 2^{-i-1}$, $l = 1$, $h_i = g = 2^{-4mn}$, $w_1 = 2^{-2mn}$, and an initial spacing of 2^{-n} for the path classes, the above recurrence shows that the path classes remain distinct. Furthermore, all the dimensions and velocities of obstacle polygons are simple rational functions of these quantities. Thus we can specify a dynamic environment which encodes a 3-SAT formula of n variables and m clauses in $O(m^2n^2)$ space and polynomial time. □

6.3 Motion Planning with Uncertainty

In our proof we construct, for any non-deterministic exponential time bounded Turing machine M, an environment that simulates that machine. We assume *wlog* that M has a binary tape (of exponential size), which initially contains its input. Given a description of M and its input, and a constant c such that the running time of M is bounded by 2^{cn}, we construct a polynomial-size environment and start and goal regions and specify an uncertainty such that *a successful motion plan exists if and only if the non-deterministic machine M has a path to an accepting state.* We do this by ensuring that any sequence of commanded motions that can possibly move the object into the goal will simulate the steps of M. Any commanded motion that does not simulate a Turing machine step prevents the object from ever being able to reach the goal. There exists a sequence of commanded motions that moves the object into the goal if and only if there is a sequence of steps of M that take it to an accepting state.

At the ith step, let $q(i$ be the internal state of M, $h(i)$ its head position, and $T(i,j)$ the contents of the jth tape square. We define the *local state* of M as the pair $\langle q(i), T(i, h(i))\rangle$.

At the start of a motion plan, the point p is somewhere in the start region S. When we execute the plan, at each time t there is a set of possible positions of p. We call this set the *instantaneous forward projection* of S and denote it $F_S(t)$. The instantaneous forward projection will consist of a number of connected components which we will call *blots.* The physical interpretation of this is that p at time t may lie anywhere inside any of the blots. We use the forward projection to represent the state of the Turing machine. The points in each blot are related by a homotopy of paths from the start region. Thus we are again making use of a free path encoding scheme, and we will be generating exponentially many blots to encode an exponential tape.

In fact, using a proof very similar to those of the previous section, we can show that verifying a single step of a motion plan with uncertainty in 3 dimensions is hard. That is:

Theorem 6.3.1 *Determining whether a point p is in the forward projection* $F_S(t)$ *at some time t for a fixed commanded velocity is NP-hard.*

6.3.1 Blot Motion

The blot model allows a simple characterization of the time evolution of the forward projection. All blots are assumed to be contained in spheres of radius $r(t)$, satisfying the following condition:

$$r(t) = \epsilon_0 + \epsilon t \tag{6.15}$$

so that the environment is initialized with all blots contained in spheres of radius ϵ_0. If a blot at time t is contained in a sphere of center $c(t)$, it should be fairly clear from (1.3) that if we command a motion v_i for time t_i, and the blot does not encounter obstacles, the blot remains spherical and its center is simply displaced to $c(t) + v_i t_i$.

If the commanded motion moves the center of the blot into a wall, then the motion of the center of the blot can be broken into two parts: a straight line motion in the direction of commanded motion, and then motion along the wall with velocity equal to the projection of the commanded velocity along the wall. In other words, the center of the blot moves compliantly as though there were no motion error.

It is easy to verify that blots satisfy the radius condition (6.15) for all motions except those where the blot is moved into a convex edge or corner. We will make use of this later to give us a splitting of blots, but generally these motions are to be avoided.

6.3.2 The Environment

Our environment can be broken into three distinct and physically separated polyhedral structures:

- Legal move filter. According to our model, it is possible to command any motion in any direction at each step. However, we would like the commanded motions to give us an orderly simulation of a Turing machine. We must therefore restrict the allowable motions to certain legal moves. The legal move filter enforces this by making it impossible to reliably reach the goal by *any* sequence of moves after an illegal move.

- Tape logic. When legal moves are executed, this structure performs the necessary updates on the tape itself, changing tape contents, head position and tape end marker fields.

- State logic. This structure performs updates on blots that represent the internal state of the Turing machine.

Initially, all of these structures contain some blots of forward projection. Now each legal move corresponds to a certain local state transition. However, the legal move filter does not examine tape contents or machine state, so at each step, the motion corresponding to every possible state transition is *legal*, even if the simulated machine is not in the state assumed by the transition. We say a legal move is *valid* if it does correspond to a Turing machine transition, given the current local state of the simulated machine. The tape and state logic structures ensure that if legal but *invalid* move is executed, then no later sequence of legal moves can reliably move the point to the goal. Thus the only possible way to reliably reach the goal is by executing a sequence of valid moves, thereby simulating Turing machine steps.

6.3.3 Legal Move Filter

The key to the simulation is to constrain *commanded* motions which may otherwise be arbitrary to a small set of motions in the x-y plane. Since the planner is free to command any motion it desires, we must build the environment in such a way that after an undesired motion, no subsequent sequence of motions can move the point to the goal. It is clear that we can constrain the motion of individual blots by placing them in a narrow channel (e.g. [Na]). Unfortunately, since we have exponentially many blots we cannot use this technique. Consider now figure 6.9. Here there are two blots in two distinct boxes, and we must command a sequence of motions that is guaranteed to move both blots into their respective goal regions. It is fairly clear that a sequence of purely horizontal motions will succeed, as long as the last motion terminates inside the goal, and there are not too many motions (because of the growth due to uncertainty). Suppose on the other hand, that we execute a single motion with a fixed vertical displacement of

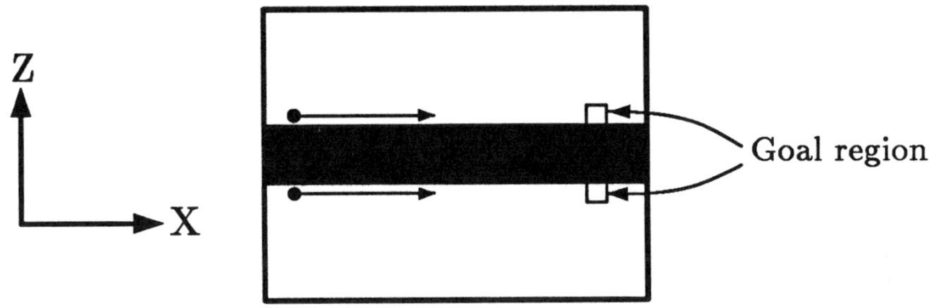

Figure 6.9: A filter for motion in the x-y plane.

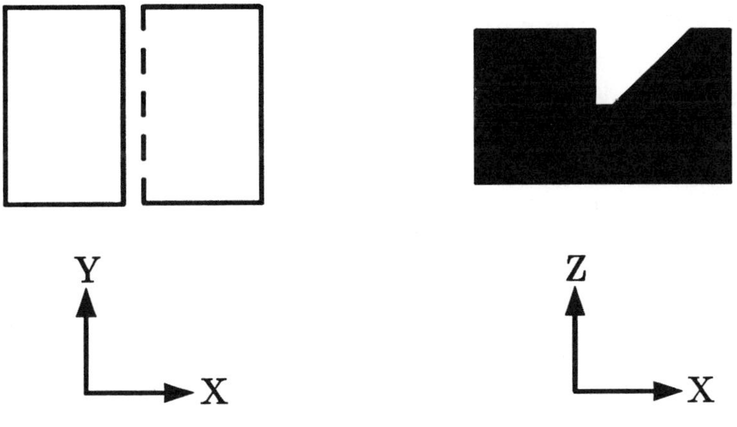

Figure 6.10: A filter for horizontal motion in the left half-plane.

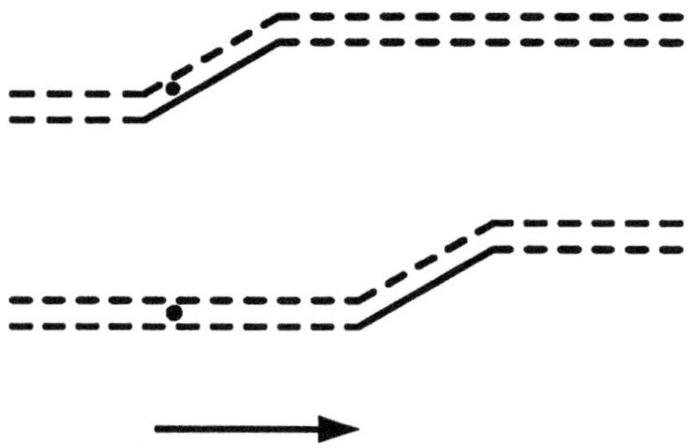

Figure 6.11: A one-way gate.

h, say upward. The bottom blot will move away from the wall by h, while the other will slide against its wall. In order to move the displaced blot to the goal, we must eventually make a motion with a downward component, but this will move the other blot away from its wall etc. Thus once a motion with a vertical displacement is executed, no sequence of motions can move both blots into the goal region. This environment functions as a filter for commanded motions, and allows only horizontal ones.

Quantitatively, if the height of the goal region is δ, and if the commanded motion causes a net vertical displacement of the center of a free blot at any time $t < t_f$ of greater than $2(\delta + \epsilon t_f)$, then from the laws of blot motion, the sum of the distances of the blot centers from the surfaces must be at least $2(\delta + \epsilon t_f)$, and can never be made less than this. It follows that one of the two blots must be entirely outside the goal at time t_f.

Once we have constrained the commanded motions to the x-y plane (or to within a region of height $2(\delta + \epsilon t_f)$), we can add a variety of other constraints. Figure 6.10 shows a structure which constrains the commanded motion directions to a semi-circle. Commanding motion into a vertical wall (shown solid) is permissible, and causes sliding in the horizontal plane, but commanding motion into a sloped wall (dashed) causes the blot to be displaced vertically. If the goal region is at the same height as the start region, the blot in this environment can never reach it once it has been displaced upward. Figure 6.11 shows a structure which contains two blots, and implements a one-way gate.

The legal move filter is shown in figure 6.12. Notice that all motions are in the x-y plane. The channels at move 3 all have one-way gates of the type shown in figure 6.11, which are not shown in figure 6.12 for simplicity. Each distinct y-level at move 3 corresponds to a particular Turing machine transition, and is indexed by the local state of M. Observe that these transitions lead to overall displacement of the tape by one square either up or down, which corresponds to the correct head motion for that transition.

Notice that none of the structures in the legal move filter have an adjacent pair of vertical walls, so that blots cannot be split in this environment. The three blots move in unison except at move 5. Three structures are necessary to ensure that blots are not separated at move 5, and that only horizontal motion to the left is possible during move 5. Notice that after a cycle of moves 1 through 5, all blots terminate in their start positions. For moves 1, 2, and 4, commanded motions may in fact be in the forward or backward direction, and this will not affect the simulation. There are one-way gates at move 3 to ensure that blots do not move in the wrong direction through tape or state logic.

We claim that the total cumulative error during a move cycle, i.e. the error between the displacement of a free blot and the distance between tape squares (take this to be 1), is a small constant times the channel width. If M runs for 2^{cn} steps (n is the input length), then we need a linear number of bits to specify ϵ and channel width small enough that the cumulative error is a small fraction of the distance between tape squares. All environment dimensions are linear in n.

6.3.4 Tape and State Logic Structures

In order that a legal move be valid, the $\langle q(i), T(i, h(i))\rangle$ pair assumed by the move must correspond to the actual local state encoded when the move is executed. For a non-deterministic machine there may be several valid moves for a given local state. The tape and state logic structures ensure that only commanded motions that simulate valid transitions can possibly lead to the goal.

Our machine has a binary tape, and the encoding of a single square

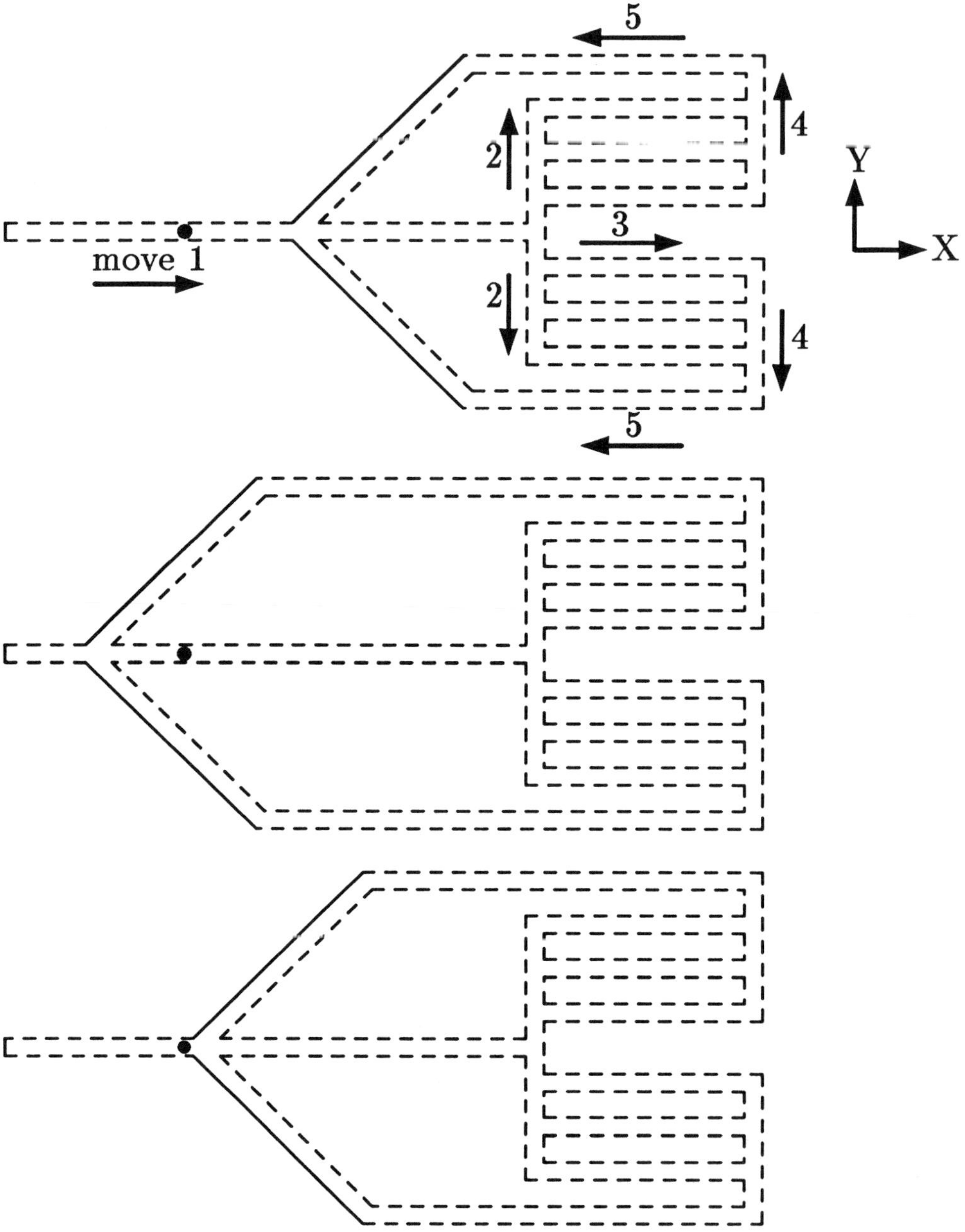

Figure 6.12: The legal move filter, consisting of three boxes.

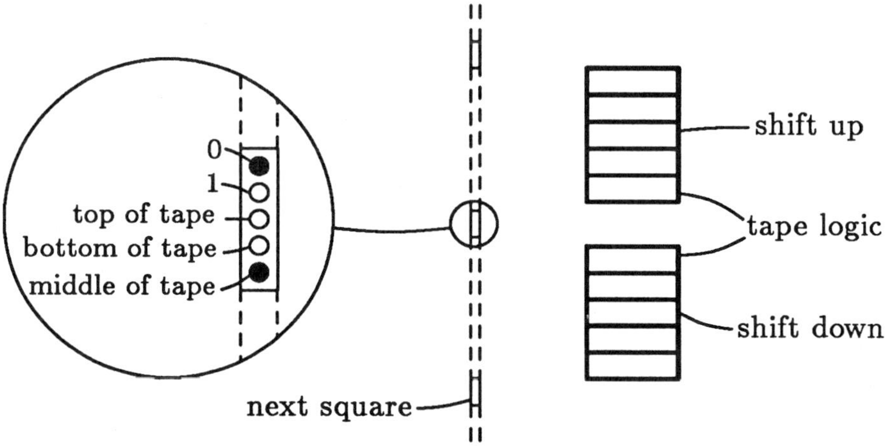

Figure 6.13: Tape encoding.

includes two rectangular regions, one to encode 0, the other 1. Exactly one of the sites must contain a blot of forward projection. Also, we have a tape end marker field with three sites, exactly one of which will contain a blot. These sites designate the square as the top, bottom or as an intermediate square. An example of an encoded tape is given in figure 6.13. Notice that the distance between tape squares is much larger than the height of each square.

Figure 6.14 shows a slice through tape and state structures at the jth level. There will be a pair of such structures for each state transition of M. Notice that there is tape logic for only one square of the tape, namely the square currently under the head. An important difference between tape and state structures is that while the tape blots are free to move up or down during the tape shift phase of a legal motion (move 4), state blots are trapped in channels and funneled back to their original y position, as shown in figure 6.15.

If the legal move corresponding to this state transition is executed but the tape contents are not $T(i,(h(i))$, then some tape blot will run into a sloped

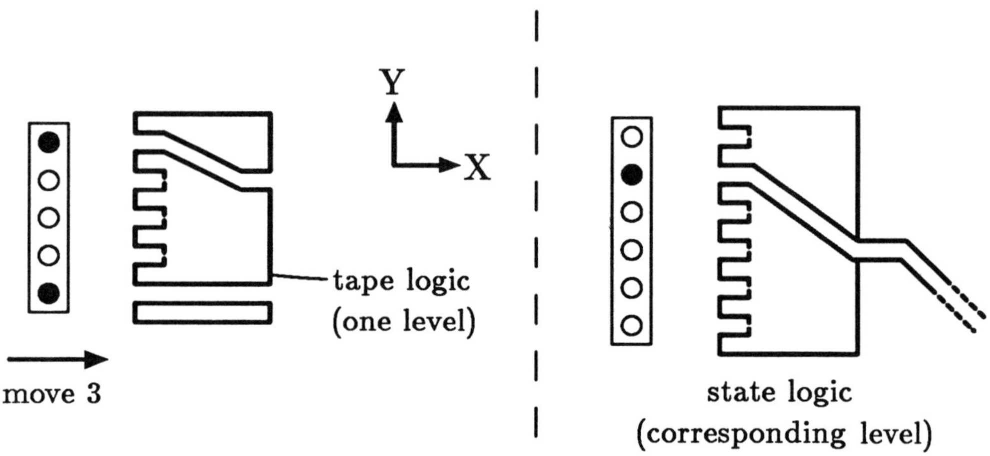

Figure 6.14: Tape and state logic (one level only).

wall and be displaced vertically. Otherwise, it will slide against a wall into the correct position for the value to be written on the tape. Similarly if the internal state before the move is not $q(i)$, the state transition structure will cause a blot to be displaced. Thus the legal move corresponding to the transition from $\langle q(i), T(i, h(i))\rangle$ will be valid if and only if the (simulated) tape contents really are $T(i, h(i))$ and the internal state is $q(i)$. Since validity of tape and state can be verified independently, tape and state structures can be physically separated.

Each level implements a state transition similar to the one described above, with the exception of transitions that require shifting beyond the limits of the tape. Such a transition is shown in figure 6.16, and denote this state transition by Q_{ij}. Here we notice that the square below the head is completely blank, but it must be correctly initialized during this move. Since there are only 2 blots entering the tape structure but we need 4 to exit, we must split some of the entering blots, as shown in the figure. Special conduits carry the blots to the location of the new tape square, since this is at some distance from the tape logic. In order for this splitting to occur,

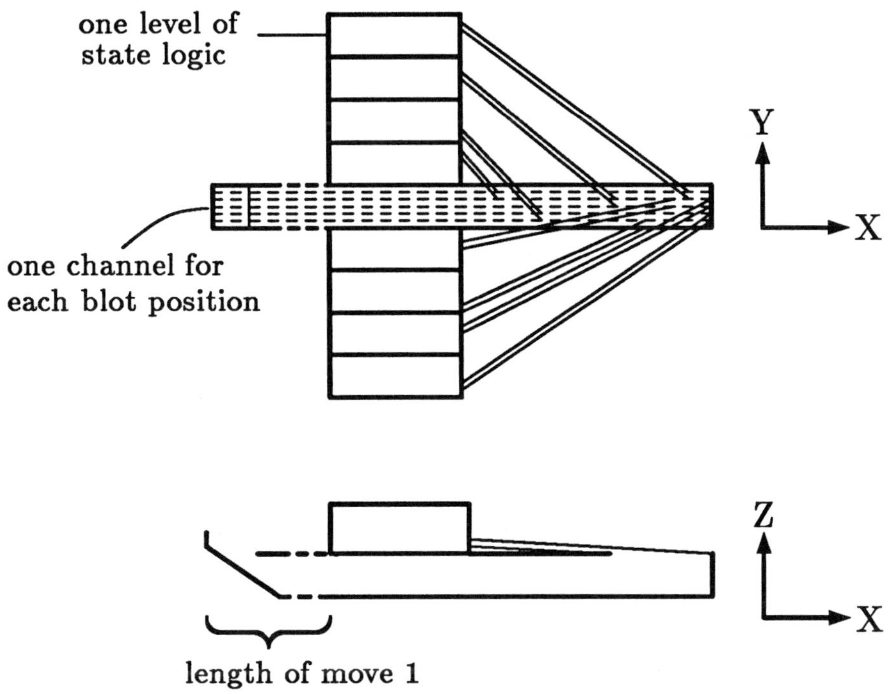

Figure 6.15: State logic (global view).

we need tighter control that normal over the direction of the commanded velocity for this move. The structure shown in figure 6.16b is added at move 3 in one of the legal move filter structures at the y-level corresponding to Q_{ij}. It ensures that some part of the forward projection must pass over both sides of the splitter in figure 6.16a.

6.3.5 Initialization and Termination

We have seen that tape and state logic structures allow only valid moves, and that any sequence of motions which reliably moves p to the goal must simulate Turing machine steps. To complete our simulation we must show that it can be correctly initialized and terminated. Initialization involves specification of the start region S, while termination involves specification

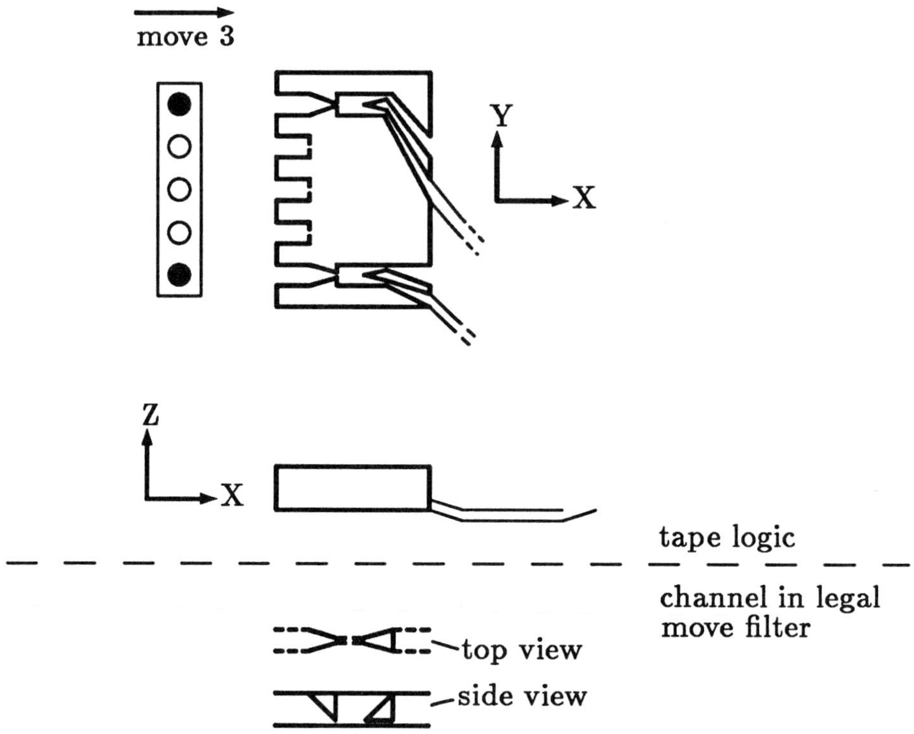

Figure 6.16: tape logic for shifting beyond the end of the tape

of the goal region G.

Firstly, the start region must correctly describe the initial internal state of the machine. This requires only a single blot (of start region) in the appropriate channel of the state logic structure. The start region must also correctly describe the initial state of the tape. This is straightforward as the tape initially needs only a linear number of blots. This is because the tape need only represent the input data to the machine M, and as explained above, if we shift beyond the tape limits, the neighboring squares are correctly setup during move 3.

The legal move generator has its blots in goal regions only at the end of move 3. The goal region for the tape structure is simply a rectangular box large enough to contain all tape blots at the end of move 3. Thus all tape blots will be in the goal region after move 3 of any sequence of valid moves. The goal region for the state logic on the other hand, is a region at the exit from the machine's halt state Q_f. At this time the point p is guaranteed to be in the goal region, and so the sequence of valid moves constitutes a succesful motion plan. Such a plan exists if and only if the non-deterministic Turing machine M halts on (accepts) that input. Thus we have

Theorem 6.3.2 *Compliant Motion Planning with Uncertainty is Non-Deterministic Exponential Time Hard.*

Proof We have described a polyhedral environment and start and goal regions such that a guaranteed motion plan exists if and only if M has a path to an accepting state. Since the amount of logic for each tape or state transition is constant, and since there is tape logic under only one square of the tape, the number of objects in the environment is polynomial in the length of the description of M. By the results of the previous sections, we need a polynomial number of bits to describe the structures and start and goal regions in the environment. Finally, the description of these structures can be computed in polynomial time. □

Chapter 7

Conclusions

Several of the chapters of this thesis describe work in relatively new areas. The treatment of these topics was far from exhaustive, and there is no shortage of open problems that follow from what has been done to date. Here we summarize these topics, and give author's selection of the most interesting unsolved problems.

7.1 Algebraic Decision Methods

In chapter 3 we presented an algorithm for computation of the multivariate resultant of a system of polynomials. The algorithm runs in single exponential sequential time, or polynomial parallel time. The multivariate resultant may be used for equation solving, via the u-resultant, or it may be used directly for variable elimination. In the former case, if a polylog parallel time method for finding approximate roots of a polynomial in one variable can be found, then equation solving would drop into polynomial parallel time. The problem of finding approximate roots fast in parallel remains open and is the most fundamental open problem in computational algebra.

There has been some progress in decision procedures for the reals and complexes, however. It turns out that it is possible to do arithmetic with real algebraic numbers fast in parallel *without numerical root approximation.* An elegant use of this idea was the paper by Ben-Or, Kozen and Reif [BKR] on decision algorithms for the full theory of the reals. Since the thesis was written, the main lemma of [BKR] on computation with algebraic numbers defined by a single polynomial has been extended by the author to systems of polynomials [Ca88a]. The generalized lemma states that it is possible to determine the signs of a collection of multivariate polynomials at the real solution points of a *system* of multivariate polynomials, and that this computation can be done in polynomial space. This generalized lemma can be used as a powerful computational subroutine, and a number of algebraic and geometric problems can be solved in PSPACE. In particular the existential theory of the reals is shown to be decidable in PSPACE in [Ca88a]. The paper also shows that the roadmap algorithm described in this thesis can be adapted to run in PSPACE, with the very satisfying consequence that the generalized movers' problem is PSPACE-complete.

The complexity of basic algebraic computations is poorly understood. Computing the multivariate resultant can be done in polynomial space, and is NP-hard, but its exact complexity is not known. Since the resultant has exponential size, a more reasonable question is to whether or not the resultant is zero, which is still NP-hard. The lower bound for the existental theory of the reals is NP-hardness, while we have only a PSPACE algorithm for it. These problems seem to be closely linked. Both look like NP problems, since if one guesses a suitable algebraic solution point, it seems plausible that one could rapidly verify that some system of equations or inequalities is satisfied or not at that point. However, there do not seem to be any compact descriptions of these algebraic witness. The tightness of the gap theorem in this thesis makes it clear that numerical approximation to witness points requires too many bits.

There is a problem with direct use of the multivariate resultant on many non-homogenous systems of polynomials. The resultant is defined only for homogeneous systems of polynomials, and non-homogeneous systems must be homogenized with an additional variable, say x_0. As we described in chapter 3, this can introduce additional solutions having $x_0 = 0$ which do not correspond to solutions of the original system. These are the solutions "at infinity". In general, the user has no control over these other solutions, and problems occur if the solutions at infinity are of higher dimension than the desired solutions. An example is the computation of the u-resultant of a system that has only finitely many affine solutions, but an infinite number of solutions at infinity. The u-resultant of such a system is identically zero. Whenever the resultant was used in this thesis, the system of polynomials had to be carefully selected to guarantee the existence of only proper solutions at infinity.

This problem has since been dealt with in [Ca88b], which describes the generalized characteristic polynomial of a system of polynomials. By symbolically perturbing the original system, it is possible to destroy the solutions of excess dimension, while only infinitesimally displacing the desired solutions. Using this idea, [Ca88b] gives a general method for recovering all the proper components of the solution set of a system of polynomials in the presence of excess dimensional components at infinity or elsewhere.

An open problem related to generalized characteristic polynomials has to do with intersection multiplicity. The degree of vanishing of the characteristic polynomial is the degree its lowest degree non-vanishing term. For characteristic polynomials of linear systems (the usual definition), the degree of vanishing counts the corank or dimension of the null space of the system. Thus it is a measure of "how much" intersection there is. One can conjecture an analagous property for generalized characteristic polynomials. Namely that the degree of vanishing counts the intersection multiplicity of the solution set. In the general case, intersection multiplicity depends not only on the solution set of the system, but on the relationship between this set and the solution sets of the individual polynomials (which are hypersurfaces). Such a result would only be generically true. Even in the linear case, the corank lemma only holds for certain systems, namely those that are diagonalizable. One application of such a result would be to efficient algorithms for the existential theory of the complex numbers. This theory is decidable in PSPACE by the results in [Ca88b] since it can be reduced to the theory of the reals. However, the multiplicity lemma would give a much more direct and efficient algorithm.

7.2 Robot Motion Planning Algorithms

We described a single-exponential time algorithm called the roadmap algorithm for skeletonizing the connected components of a semi-algebraic set. For this result we used stratified sets to partition the semi-algebraic set into smooth pieces. Stratified sets seem a natural abstraction for representing algebraic and semi-algebraic sets geometrically, rather than their defining equations. The stratification of a semi-algebraic set may be thought of as a kind of boundary representation to use the CAD terminology. In fact the roadmap of a semi-algebraic set contains the roadmaps of the closures of all the strata in the set. Adjacency information can be obtained from the curves, so that there is sufficient information in the generic case in the roadmap to construct the adjacency graph of the stratification. The roadmap algorithm may therefore be viewed as a kind of decomposition algorithm.

While adjacency information for strata is useful, it is not sufficient to characterize the topology of the semi-algebraic set. Each stratum may have complicated topological structure, and we have no information on this. To characterize the set topologically, we need a triangulation or cell decomposition. However, it does seem likely that the techniques in this thesis may be extended to compute a coarse cell decomposition or triangulation in single exponential time.

Such a decomposition could be used for point queries in semi-algebraic sets. Given a point $p \in \Re^r$, can we determine whether it lies in a given semi-algebraic set S without evaluating all the polynomials that define S ? Collins decomposition allows queries in time that grows logarithmically with the number of polynomials, but double exponentially with the number of variables.

The euclidean shortest path and asteroid avoidance problem can both be polynomial-time reduced to problems in the existential theory of the reals. So by the results of [Ca88a] they can be solved in PSPACE. These problems are only the first steps toward a general motion planner. Much more work needs to be done to deal with more complicated dynamic constraints, such as limits on accelerations, and more complicated cost functions, that depend on time taken or perhaps energy used. Solutions to these problems will require new techniques for path representation. We have seen that the generalized movers' problem can be solved using algebraic paths if the obstacles have algebraic boundaries. But trajectories that satisfy algebraic dynamic constraints are not algebraic in general. The results of chapter 4 are still somewhat relevant at least, since the roadmap definition and most of its properties do not require the set S to be semi-algebraic, only stratifiable. The notion of stratifications, which represent equivalence classes of points, will need to be augmented with the notion of foliations, representing sets of trajectories which satisfy some dynamic equation.

A significant number of important planning problems seem to fall into a qualitatively more difficult class, even though they seem very similar the problems we have discussed. These problems involve *non-holonomic* constraints on motion. Non-holonomic constraints are those where the set of infinitesimal motions that the object can perform (in its configuration space) is not determined solely by its configuration. Strictly speaking, the generalized movers' problem involves non-holnomic constraints, but they are still piece-wise holonomic, (holonomic within each stratum of the set of free configurations) and are much simpler to deal with than general non-holonomic constraints. A simple example of a non-holonomic constraint is that which keeps an ice skater from instantaneously changing direction, or in going in any direction other the direction in which his skates are pointed. Examples that arise in robotics include wheeled vehicles, which are very similar to the ice skater, and have a fixed lower bound on their turning radius. The problem of finding a bounded turning radius path for a point in two dimensions is addressed in [FW] and is there shown to be solvable in exponential time. Planning problems with dynamic constraints are non-holonomic, and similar techniques may be applicable to both types of problem. In particular, it is necessary to represent sets of equivalent trajectories, and again foliations seem to be the right mathematical tool.

Some of the ideas from the roadmap algorithm may be applicable to collision-free grasp planning. Sophisticated robot hands have large numbers of degrees of freedom (≥ 12), which seems to be out of the realm of feasibility for exact algorithms like the roadmap algorithm. On the other hand, the configuration space of a hand has very special structure. Specifically, if the hand has k fingers, the configuration space is a k-fold product of the space for one finger. If the complexity of the one-finger configuration space can be kept low, the hand configuration space has an efficient implicit representation over it. Another interesting possibility is constructive stratification of the various jacobians between fingers and the object (smooth real-valued functions, like manifolds, can be stratified). The different strata correspond to grasps having qualitatively different properties. Knowing the connectivity of these strata allows one to determine if the jacobian can be continuously changed (corresponding to continuous motion of the finger contact points) from one grasp to another while always maintaining certain properties of the grasp.

Much work remains to be done on the problem of motion planning with uncertainty. In this thesis we gave only a negative result, showing that the worst case complexity is extremely high (non-deterministic exponential time hard) if an unbounded number of steps is allowed. On the theoretical side, there are no upper bounds on the problem whatsoever, and it is not even clear that it is decidable. In practice, any realistic application will involve only a small number of steps, and our proofs are "unstable" numerically. That is, a small perturbation of the environment destroys the proof, and apparently simplifies the problem dramatically. So the problem may still be amenable to an exact algorithm, but one whose complexity depends on the complexity of the problem instance, not just its size. Even the difficult example we produced may well be solvable in time polynomial in the *output size*, which may also be polynomial in the *input size* with probability close to one.

7.3 Lower Bounds

In chapter 6 we used free path encoding to give lower bounds for three fundamental robot motion planning problems. We gave the first non-trivial lower bounds for 3-dimensional shortest path, and our proof applies to any L^p metric. We showed that finding the shortest path is NP-hard, and that a polyhedral environment may contain exponentially many shortest path classes. The upper bound for the problem is now polynomial space, but the exact complexity remains open. The difficulty in finding good bounds seems to stem from the fact that the problem has both a combinatorial (edge sequence touched) and an algebraic character (position of contact points on edges). We made use of the combinatorial aspect only in our NP-hardness proof.

However, knowing the edge sequence is not sufficient to determine the path, because the position of contact points on edges does not seem to be describable as a polynomial-size algebraic number, so there is no obvious NP or Σ_2 algorithm for the problem. The contact points are defined by high degree polynomials, and Bajaj [Ba] has shown these polynomials are irreducible in some cases. This may be viewed as a special case of the general problem we mentioned earlier of finding good lower bounds on problems that depend on algebraic computation with many variables.

Notice that our proof requires some large rational numbers to describe the environment, and no longer succeeds if integers are coded in unary. Thus we have demonstrated NP-hardness but not strong NP-hardness. This is not surprising in the light of Papadimitriou's [Pa] approximate algorithm for the problem, which takes time polynomial in both the environment size and the length of a unary approximation error. However, our proof does show that determining a polynomial number of bits of the length of the shortest path is difficult, so that better approximate methods do not look feasible. One direction that may be fruitful is consideration of restricted classes of environments, for example, where all the obstacles are rectilinear. Our proof depends crucially on the existence of diagonal edges to allow general placement of virtual sources. If these edges are removed, the reflection of virtual sources about obstacle edges is drastically simplifed, and characterizing this may well lead to an efficient algorithm.

We showed that 2-dimensional dynamic motion planning with bounded velocity magnitude is NP-hard even with polygonal (in fact convex) obstacles moving with constant velocity. This result should readily generalize to the case where the velocity is bounded by any convex polygon, rather than a circle.

We also gave a non-deterministic exponential time hardness proof for the motion planning with uncertainty problem. Here control uncertainty was a fundamental limitation to our representation of state using path encoding. It seems plausible that the proof could be extended to exponential space hardness by periodically squeezing all of the state blots, with a "refresh" of the tape.

References

[Baj] Bajaj C., "The Algebraic Complexity of Shortest Paths in Polyhedral Spaces", Purdue University, Computer Science tech. rept. CSD-TR-523, (June 1985).

[BKR] Ben-Or M., Kozen D., and Reif J., "The Complexity of Elementary Algebra and Geometry", J. Comp. and Sys. Sciences, Vol. 32, (1986), pp. 251-264.

[Ber] Berman L., "Precise Bounds for Presburger Arithmetic and the Reals with Addition", Proc. 18th IEEE Symp. FOCS, (1977), pp. 95-99.

[BL] Brooks R. A. and Lozano-Pérez T., "A Subdivision Algorithm in Configuration Space for Findpath with Rotation", IJCAI 83, pp. 799-806.

[Boy] Boyse J. W., "Interference Detection Among Solids and Surfaces", Comm ACM, vol 22, No 1 (1979) pp. 3-9.

[BS] Bayer D. and Stillman M., "On the Complexity of Computing Syzygies", manuscript, Harvard University (1985).

[Buc65] Buchberger B., "An algorithm for finding a basis for the residue class ring of a zero-dimensional polynomial ideal", (in German) Ph.D. thesis, Univ. of Innsbruck, Austria (1965).

[Buc85] Buchberger B., "Gröbner: An algorithmic method in polynomial ideal thoery", in chapter 6 of "Multidimensional systems theory", N. K. Bose ed., D Reidel Publishing Co. (1985).

[Buc87] Buckley S., "Planning and Teaching Compliant Motion Strategies", MIT AI Lab TR-936, (1987).

[Cam] Cameron S., "A Study of the Clash Detection Problem in Robotics", proc. IEEE conf. on Robotics and Automation, (1985), pp. 488-493.

[Ca85] Canny J. F., "A Voronoi Method for the Piano-Movers Problem", Proc. IEEE Int. conf. Robotics and Automation, (March 1985), pp. 530-535.

[Ca86] Canny J. F., "Collision Detection for Moving Polyhedra", IEEE trans. PAMI, vol 8, no 2, (March 1986), pp. 200-209.

[Ca87] Canny J. F. "Constructing Roadmaps of Semi-Algebraic Sets", Proc. Symp. on Geometric Reasoning, Oxford University (to be published 1988).

[Ca88a] Canny J. F. "Some Algebraic and Geometric Computations in PSPACE", Proc. 20th ACM Symp. Theory of Computing, Chicago, (1988).

[Ca88b] Canny J. F., "Generalized Characteristic Polynomials", submitted to ISSAC-88/AAECC-6, Rome, (July 1988).

[Cha] Chazelle B., "Convex Partitions of Polyhedra", SIAM Jour. Computing, Vol. 13, No. 3, (1984), pp 488-507.

[CG] Chistov A. L. and Grigoryev D. Y., "Complexity of quantifier elimination in the theory of algebraically closed fields", Lect. Notes Comp. Sci. 176, Springer Verlag, (1984).

[CL] Collins G. E. and Loos R., "Real Zeros of Polynomials", Computing Supplementum 4, (1982) pp. 83-94.

[Coh] Cohen P. J., "Decision Procedures for Real and p-adic Fields", comm. Pure and App. Math. vol XXII, no. 2, (March 1969), pp. 131-151.

[Col] Collins G. E. "Quantifier Elimination for Real Closed Fields by Cylindrical Algebraic Decomposition", Lecture Notes in Computer Science, No. 33, Springer-Verlag, New York, (1975), pp. 135-183.

[Csa] Csanky L., "Fast Parallel Matrix Inversion Algorithms", SIAM J. Comp., Vol. 5, No. 4, (Dec. 1976), pp. 618-623.

[DH] Denavit J., and Hartenberg R. S., "A Kinematic Notation for Lower-Pair Mechanisms Based on Matrics", ASME Jour. of Applied Mechanics, (June 1955), pp. 215-221.

[DGHS] Dobkin D., Guibas L., Hershberger J., Snoeyink J., "An efficient algorithm for finding the CSG representation of a simple polygon", to appear in SIGRAPH, (1988).

[Don] Donald, B. R., "Motion Planning with Six Degrees of Freedom", MIT AI-TR-791, MIT Artificial Intelligence Lab., (1984).

[EOS] Edelsbrunner H., O'Rourke J., Seidel R., "Constructing Arrangements of Lines and Hyper-planes with Applications", Proc. 24th Symp. on Foundations of Computer Science, (1983), pp. 83-91.

[Erd] Erdmann M., "On Motion Planning with Uncertainty", MIT AI Lab TR-810 (1984).

[FW] Fortune S., and Wilfong G., "Planning Constrained Motions", Proc. 20th ACM Symp. Theory of Computing, Chicago, (1988).

[GG] Golubitsky M. and Guilleman V., "Stable Mappings and Their Singularities", GTM-14, Springer Verlag, New York, (1973).

[GP] Guilleman V. and Pollack A., "Differential Topology", Prentice-Hall Inc., Englewood Cliffs, New Jersey (1974).

[Gri] Grigoryev D. Y., "Complexity of Deciding Tarski Algebra", Jour. Symbolic Computation, special issue on decision algorithms for the theory of real closed fields, to appear (1987).

[GV] Grigoryev D. Y., and Vorobjov N. N. (Jr.), "Solving Systems of Polynomial Inequalities in Subexponential Time", op. cit. (1987).

[GWDL] Gibson C. G., Wirthmüller K., Du Plessis A. A., Looijenga E. J. N., "Topological Stability of Smooth Mappings", Lecture Notes in Mathematics, No. 552, Springer-Verlag, New York, (1976).

[Ham] Hamilton W. R., "Elements of Quaternions", 3rd edition, Chelsea Publishing Co., New York, (1969).

[Hir] Hironaka H., "Resolution of Singularities of an Algebraic Variety over a Field of Characteristic Zero", Ann. of Math. 79, (1964) pp. 109-326.

[Hon] Hong J. W., "Proving by Example and Gap Theorem", Proc. 27th IEEE Symp. FOCS, (1986), pp. 107-116.

[HSS] Hopcroft J., Schwartz J., Sharir M., "On the Complexity of Motion Planning for Independent Objects: PSPACE-Hardness of the 'Warehouseman's Problem' ", Int. Jour. Robotics Research, vol 3, no 4, Winter (1984), pp. 76-88.

[Hur] Hurwitz A. "Über die Trägheitsformen eines algebraischen Moduls", Annali di Mat., Ser. III, Tomo XX, (1913), pp. 113-151.

[HW] Hopcroft J., and Wilfong G., "Motion of Objects in Contact", Int. Jour. Robotics Res. vol 4, no. 4, (1986), pp. 32-46.

[Ino] Inoue H., "Force Feedback in Precise Assembly Tasks", MIT AI Lab memo 308, (August 1974).

[Joh] Johnson F. E. A., "On the Triangulation of Stratified Sets and Singular Varieties", Trans. Amer. Math. Soc. 275, (1983), pp. 333-343.

[Kal] Kaltofen E., "A polynomial-time reduction from bivariate to univariate integral polynomial factorization", Proc. 23rd IEEE Symp. Foundations of Comp. Sci., (1982), pp. 57-64.

[KY] Kozen D., and Yap C. "Algebraic Cell Decomposition in NC", Proc IEEE symp. FOCS, (1985), pp. 515-521.

[Laz81] Lazard D., "Résolution des Systèmes d'Équations Algébriques", Theor. Comp. Sci. vol 15, (1981), pp. 77-110.

[Laz83] Lazard D., "Gröbner Bases, Gaussian Elimination and Resolution of Systems of Algebraic Equations", Proc. Eurocal 83 conf., London (1983), pp. 146-156.

[LD] Lee, D. T., and Drysdale, R. L., "Generalization of Voronoi diagrams in the plane", SIAM J. Comp. (10) (1981) pp. 73-87.

[LMT] Lozano-Pérez T., Mason M. T., and Taylor R. H., "Automatic Synthesis of Fine Motion Strategies for Robots", Int. Jour. Robotics Research, vol 3, no 1, (Spring 1984), pp. 3-24.

[Loz] Lozano-Pérez T., "Spatial Planning: A Configuration Space Approach", IEEE Trans. Computers, C-32, No. 2 (Feb 1983) pp. 108-120.

[LW] Lozano-Pérez T., and Wesley M., "An algorithm for Planning Collision-Free Paths Among Polyhedral Obstacles", Comm. ACM, vol 22, no 10, (Oct 1979), pp. 560-570.

[Mac] Macaulay F. S., "Some Formulae in Elimination", Proc. London Math. Soc. (1) 35 (1902) pp. 3-27.

[Mas] Mason M., "Compliance and Force Control for Computer Controlled Manipulators", IEEE Trans. SMC, vol 11, no 6, (June 1981), pp. 418-432.

[Mer] Mertens F. "Über die bestimmenden Eigenschaften der Resultante von n Formen mit n Veränderlichen", Sitzungsberichte der Kais. Akad. d. Wissenschaften zu Wien, Bd. 93 (1886).

[Mig] Mignotte M., "Some Useful Bounds", in "Computer Algebra, Symbolic and Algebraic Computation", Buchberger et al. ed., Springer-Verlag, New York, (1982), pp. 259-263.

[Mil63] Milnor J., "Morse Theory", Annals of Math. series, No. 51, Princeton University Press, Princeton, New Jersey, (1963).

[Mil64] Milnor J., "On the Betti Numbers of Real Varieties", Proc. Amer. Math. Soc. 15, (1964), pp. 275-280.

[MM] Mayr E., Meyer A., "The Complexity of the Word Problem for Commutative Semigroups and Polynomial Ideals", Advances in Math. 46, (1982), pp. 305-329.

[Mou] Mount D. M., "On finding shortest paths on convex polyhedra", Tech. Rept., Computer Science Department, University of Maryland, (1984).

[Mum] Mumford D., "Algebraic Geometry I, Complex Projective Varieties", Springer-Verlag, New York, (1976).

[Nat] Natarajan B. K., "On Moving and Orienting Objects", Cornell University, TR86-775, (1986).

[OSYa] Ó'Dúnlaing, C., Sharir, M., and Yap C., "Generalized Voronoi diagrams for moving a ladder: I Topological Analysis", NYU-Courant Institute, Robotics Lab. Tech. report No. 32 (1984).

[OSYb] Ó'Dúnlaing, C., Sharir, M., and Yap C., "Generalized Voronoi diagrams for moving a ladder: II Efficient construction of the diagram", NYU-Courant Institute, Robotics Lab. Tech. report No. 33 (1984).

[OY] Ó'Dúnlaing C., and Yap C., "A retraction method for planning the motion of a disc", J. Algorithms (6), (1985), pp. 104-111.

[Pap] Papadimitriou C., "An Algorithm for Shortest-Path Motion in Three Dimensions", Inf. Proc. Letters, vol 20, (1985), pp. 259-263.

[Pau] Paul R. P., "Robot Manipulators", MIT Press, Cambridge Mass., (1981).

[PW] Pervin E. and Webb J. "Quaternions in Computer Vision and Robotics", CMU report CS-82-150, (1982).

[Pri] Prill D., "On Approximations and Incidence in Cylindrical Algebraic Decompositions", SIAM Jour. Comp., Vol. 15, No. 4, (Nov. 1986), pp. 972-993.

[Rei] Reif J., "Complexity of the Mover's Problem and Generalizations", Proc. 20th IEEE Symp. FOCS, (1979). Also in "Planning, Geometry and Complexity of Robot Motion", ed. by J. Schwartz, J. Hopcroft and M. Sharir, , Ablex publishing corp. New Jersey, (1987), Ch. 11, pp. 267-281.

[RSh] Reif J., and Sharir M., "Motion Planning in the Presence of Moving Obstacles", Proc. 25th IEEE symp. FOCS, (1985), pp. 144-154.

[RSt] Reif J., and Storer J., "Shortest Paths in Euclidean Space with Polyhedral Obstacles", Tech. Rep. CS-85-121, Comp. Sci. Dept., Brandeis University, (April 1985).

[Ren] Renegar J., "On the Worst-Case Arithmetic Complexity of Approximating Zeros of Systems of Polynomials", Technical Report, School of Operations Research and Industrial Engineering, Cornell University, (May 1987).

[Sa] Salamin E., "Application of Quaternions to Computation with Rotations", Stanford AI Lab Internal working paper, (1979).

[Sal] Salmon G., "Modern Higher Algebra", G. E. Stechert and Co., New York, reprinted 1924, (1885).

[Sei] Seidenberg A., "Constructions in Algebra", trans. AMS 197, (1974) pp. 273-313.

[SHS] Schwartz J., Hopcroft J., and Sharir M. (eds.) "Planning, Geometry and Complexity of Robot Motion Planning", Albex Publishing Co., New Jersey, (1987).

[SSh] Schwartz J. and Sharir M., "On the 'Piano Movers' Problem, II. General Techniques for Computing Topological Properties of Real Algebraic Manifolds", Comp. Sci. Dept., New York University report 41, (1982). Also in "Planning, Geometry and Complexity of Robot Motion", ed. by J. Schwartz, J. Hopcroft and M. Sharir, Ablex publishing corp. New Jersey, (1987), Ch. 5, pp. 154-186.

[SY] Schwartz J. and Yap C. K., "Advances in Robotics", Lawrence Erlbaum associates, Hillside New Jersey, (1986).

[SB] Sharir M., and Baltsam A., "On Shortest Paths amid Convex Polyhedra", Proc. ACM Computational Geometry Conf., Yorketown Heights, (1986), pp. 193-206.

[SSc] Sharir M., and Schorr A., "On Shortest Paths in Polyhedral Spaces", Proc. 16th ACM STOC, (1984), pp. 144-153.

[Tar] Tarski A., "A Decision Method for Elementary Algebra and Geometry", Univ. of Calif. Press, Berkeley, (1948), second ed. 1951.

[Th65] Thom R., "Sur L'Homologie des Variétés Algébriques Réelles", in "Differential and Combinatorial Topology", S. Cairns ed., Princeton University Press, New Jersey, (1965), pp. 255-265.

[Th69] Thom R., "Ensembles et Morphismes Stratifiés", Bull. AMS. vol 75, (1969), pp. 240-284.

[Udu] Udupa S., "Collision Detection and Avoidance in Computer Controlled Manipulators", Proc. 5th Int. Joint. Conf. on Art. Intell., Mass. Inst. Tech. (1977), pp. 737-748.

[Wae] van der Waerden B. L., "Modern Algebra", (third edition) F. Ungar Publishing Co., New York (1950).

[Wal] Wall C. T. C., "Regular Stratifications", in "Dynamical Systems - Warwick 1974", Springer Lect. Notes in Math. No. 468, (1974) pp. 332-344.

[Wh57] Whitney H., "Elementary Structure of Real Algebraic Varieties", Annals of Math., Vol. 66, No. 3, (Nov. 1957), pp. 545-556.

[Wh65] Whitney H., "Tangents to an Analytic Variety", Annals of Math. 81, (1965), pp. 496-549.

[Wu] Wu W. T., "Basic Principles of Mechanical Theorem Proving in Elementary Geometry", Jour. Sys. Sci. Math. Sci. 4, (1984), pp. 207-235.

[Yan] Yang A. T., "Applications of Quaternion Algebra and Dual Numbers to the Analysis of Spatial Mechanisms", University Microfilms, Inc., Ann Arbor, Michigan, (1964).

[Yap] Yap, C. K., "Coordinating the motion of several discs", NYU-Courant Institute, Robotics Lab. Rept. No. 16 (1984).

Index

The MIT Press, with Peter Denning, general consulting editor, and Brian Randell, European consulting editor, publishes computer science books in the following series:

ACM Doctoral Dissertation Award and Distinguished Dissertation Series

Artificial Intelligence, Patrick Winston and Michael Brady, editors

Charles Babbage Institute Reprint Series for the History of Computing, Martin Campbell-Kelly, editor

Computer Systems, Herb Schwetman, editor

Exploring with Logo, E. Paul Goldenberg, editor

Foundations of Computing, Michael Garey and Albert Meyer, editors

History of Computing, I. Bernard Cohen and William Aspray, editors

Information Systems, Michael Lesk, editor

Logic Programming, Ehud Shapiro, editor; Fernando Pereira, Koichi Furukawa, and D. H. D. Warren, associate editors

The MIT Electrical Engineering and Computer Science Series

Scientific Computation, Dennis Gannon, editor

Hey Mike!

I'd be honored to be on your math bookshelf. Many thanks

-John